Niko Tzoukmanis

Local Minimizers
of Singularly Perturbed Functionals
with Nonlocal Term

λογος

Augsburger Schriften zur Mathematik, Physik und Informatik
Band 7

herausgegeben von:
Professor Dr. F. Pukelsheim
Professor Dr. B. Aulbach
Professor Dr. W. Reif
Professor Dr. D. Vollhardt

Bibliografische Information Der Deutschen Bibliothek

Die Deutsche Bibliothek verzeichnet diese Publikation in der Deutschen
Nationalbibliografie; detaillierte bibliografische Daten sind im Internet über
http://dnb.ddb.de abrufbar.

ISBN 3-8325-0650-0
ISSN 1611-4256

Logos Verlag Berlin
Comeniushof, Gubener Str. 47,
10243 Berlin
Tel.: +49 030 42 85 10 90
Fax: +49 030 42 85 10 92
INTERNET: http://www.logos-verlag.de

Vorwort

Bei Phasenübergängen im festen Zustand werden in gewissen Materialien zwei Zustände beobachtet: Eine homogene Phase, genannt Austenit, und eine zweite Phase, welche nicht homogen ist, sondern in ihrer kristallinen Struktur lange Lamellen, sogenannte Zwillinge ("twins"), ausbildet, welche als Martensit bezeichnet wird. Die gleichzeitige Ausbildung beweist, dass keine Phase energetisch bevorzugt wird. Für die mathematische Modellierung bedeutet dies, dass das Potential der elastischen Energie zwei Minima auf gleicher Höhe besitzt. Es ist ein sogenanntes Zweitopfpotential, welches offensichtlich nicht konvex ist. Ziel der Modellierung ist es, die charakteristische Periode der Zwillinge, d.h. die Skala der kristallinen Mikrostruktur des Martensit, zu bestimmen.

Als zufriedenstellend erweist sich ein von Kohn und Müller im Jahr 1992 entwickeltes zweidimensionales Modell, aus welchem die Energie entlang der Grenze, dem "interface", zwischen Austenit und Martensit bestimmt werden kann. Dies ist folglich ein eindimensionales Modell mit einem Zweitopfpotential und einem Spurterm, welcher durch Interpolation zwischen der 0. und der 1. Ableitung entsteht. Dieses Quadrat der sogenannten $\frac{1}{2}$-Norm ist ein nichtlokaler Term mit der unangenehmen Eigenschaft, dass er sich bezüglich disjunkter Intervalle nicht additiv verhält. Das Problem, diese elastische Energie des "interface" zu minimieren, ist schon deshalb schwierig, da sie wegen der Nichtkonvexität nicht unterhalb stetig ist, und deshalb die direkten Methoden der Variationsrechnung nicht anwendbar sind. Im vorliegenden Fall kann man in der Tat zeigen, dass kein Minimierer existiert, das Modell demnach aus physikalischer Sicht noch nicht zufriedenstellend ist.

Aus diesem Grund haben Kohn und Müller eine konvexe Relaxation durch das Quadrat der 2. Ableitung hinzugefügt — eine sogenannte singuläre Störung —, welche nicht nur mathematisch sondern auch physikalisch sinnvoll ist: Dieser zusätzliche Term beschreibt die Oberflächenenergie des Martensit. S. Müller hat 1993 in einer viel beachteten Arbeit gezeigt, dass die Größenordnung ε der singulären Störung die Periode des Minimierers und damit die Mikrostruktur des Martensit skaliert, sofern das Quadrat der $\frac{1}{2}$-Norm durch das Quadrat der lokalen 0-Norm ($= L^2$-Norm) ersetzt wird. Er hat vermutet, dass dies auch für die Energie mit der $\frac{1}{2}$-Norm richtig bleibt, allerdings mit einer unterschiedlichen Skalierung in ε. Diese Vermutung führte zu der Themenstellung der Dissertation von Herrn Tzoukmanis.

Bar jeder Physik stellt sich also das Problem, ein singulär gestörtes Funktional 2. Ordnung mit einem Zweitopfpotential 1. Ordnung und einem nichtlokalen Term der Ordnung $\frac{1}{2}$ über einem Intervall zu minimieren. Wegen der Konvexität der singulären Störung und der Koerzivität des Funktionals ist die Existenz eines Minimierers klar. Das Problem ist, seine Struktur zu bestimmen, wenn die singuläre Störung mit ε gegen Null konvergiert.

Zu diesem Zweck hat De Giorgi Mitte der 70iger Jahre die Γ-Konvergenz erfunden: Ein Funktional heißt Γ-Grenzwert von Funktionalen, falls, grob gesprochen, deren Minimierer gegen einen Minimierer konvergieren. Auf diese Weise erhält man aus dem Minimierer des Γ-Grenzwertes Informationen über die Minimierer der approximierenden Funktionale, was sinnvoll ist, wenn der Γ-Grenzwert eine wesentliche Vereinfachung des Problems darstellt. Im vorliegenden Fall ist der Γ-Grenzwert, wenn die singuläre Störung gegen Null konvergiert, im Wesentlichen von Modica und Mortola im Jahre 1977 gegeben worden: Es ist die Summe eines diskreten Funktionals und der nichtlokalen $\frac{1}{2}$-Norm, welche nur auf periodische Sägezahnfunktionen mit Steigungen ± 1 wirkt, wenn bei ± 1 die Minima des Zweitopfpotentials liegen. Der diskrete Term zählt die Zähne pro Periode. Für den vorliegenden Fall musste Herr Tzoukmanis das Resultat von Modica-Mortola allerdings auf den periodischen und translationsinvarianten Fall verallgemeinern (Minimierer treten als Orbits auf).

Da das diskrete Zählfunktional lokal konstant ist (in der dieser Theorie zugrunde liegenden Topologie), besteht das letzte Problem darin, die nichtlokale $\frac{1}{2}$-Norm auf der Klasse der periodischen Sägezahnfunktionen mit einer vorgegebenen Anzahl von Zähnen zu minimieren. Das klingt einfach und man erwartet, dass der Minimierer eine regelmäßige Säge mit äquidistanten Zähnen ist. Ist dies der Fall, so sind nach De Giorgi die Minimierer der singulären Störungen für kleine ε (also des physikalischen Energiefunktionals) ebenfalls Funktionen, welche in der Nähe solcher regelmäßigen Sägezahnfunktionen liegen und somit fast periodisch sind.

In einer Arbeit von Alberti und Müller aus dem Jahr 2001 wird diese Vermutung geäußert. Die Tatsache, dass kein Beweis gegeben wurde, zeigt, dass dieser wohl nicht auf der Hand lag.

In der Tat, das nichtlokale Funktional gegeben durch die $\frac{1}{2}$-Norm ist selbst auf der Klasse der Sägezahnfunktionen so unübersichtlich, dass alle Versuche, die äquidistanten Sägen als (lokale) Minimierer nachzuweisen, scheiterten. Herrn Tzoukmanis gelang der Durchbruch allerdings nur für eine Klasse von nichtlokalen Funktionalen, welche die $\frac{1}{2}$-Norm in geeigneter Weise approximieren: Während diese eine Faltung mit unbeschränktem Kern ist, haben seine zulässigen Funktionale beschränkte Kerne. (Ich habe gehört, dass diese Funktionale die Physik sogar besser beschreiben.)

Für solche nichtlokalen Funktionale lautet das Hauptergebnis wie folgt: Regelmäßige Sägen mit äquidistanten Zähnen sind lokale Minimierer, sofern die Anzahl der Zähne hinreichend groß ist. Die Schlussfolgerung für lokale Minimierer der physikalischen singulär gestörten Funktionale wurde bereits erwähnt.

Der Beweis dieses einfach zu formulierenden Resultats ist allerdings äußerst verwickelt und verlangt viel Technik. Der Grund liegt darin, dass kleine lokale Variationen von (Sägezahn-) Funktionen nicht lokalisierbar sind, sondern durch die Faltung über das gesamte Intervall "verschmieren". Dies zu kontrollieren und quantitativ zu fassen, ist sehr schwierig.

Die Vorgehensweise ist klar: Die 1. Variation muss verschwinden, die 2. Variation muss eine positiv definite Bilinearform sein. Letztere wird (auf der Klasse der Sägezahnfunktionen) durch eine symmetrische Matrix dargestellt, welche eine spezielle Struktur haben und als "zyklische Matrizen" bezeichnet werden. (Es sind spezielle Toeplitz-Matrizen.) Dies eröffnet die Möglichkeit, die Eigenwerte explizit anzugeben und deren Positivität zu beweisen. Die Rechnung basiert auf einer geeigneten additiven Zerlegung der Toeplitz-Matrizen in drei Summanden, deren ersterer positive Eigenwerte besitzt. Der Rest ist eine äußerst komplizierte Störungstheorie, die zeigt, dass asymptotisch bei großen Dimensionen die Positivität erhalten bleibt.

Augsburg, den 30. Juni 2004

Prof. Dr. Hansjörg Kielhöfer
Institut für Mathematik
Mathematisch-Naturwissenschaftliche Fakultät
Universität Augsburg

Abstract:

We study a nonlocal variational problem which models the interface of an austenite/twinned martensite interface. We derive the limit functional via the Modica-Mortola theorem and describe the connection between its local minimizers and those of the original functionals. The limit functional is only finite on a set of sawtooth functions. We will show that for N large, a sawtooth with N equidistant corners is a local minimizer of the nonlocal energy and thus of the limit functional. This property is connected to a positivity condition for a circulant matrix. Its eigenvalues, which are given explicitly through the matrix coefficients, are estimated.

From the local minimizing property for the nonlocal energy, we will finally deduce the main result, which is the existence of strongly oscillating near-periodic local minimizers of the original functionals.

Zusammenfassung:

Wir betrachten ein nichtlokales Variationsproblem, welches ein Interface zwischen Austenit und Martensit modelliert. Mittels des Satzes von Modica-Mortola leiten wir das Grenzfunktional her und werden den Zusammenhang zwischen den lokalen Minimierern desselben und denen der ursprünglichen Funktionale beschreiben. Das Grenzfunktional ist nur auf einer Menge von Sägezahnfunktionen endlich. Wir werden zeigen, dass eine Sägezahnfunktion mit N äquidistanten Ecken ein lokaler Minimierer der nichtlokalen Energie und damit auch des Grenzfunktionals ist, sofern N groß genug ist. Diese lokale Minimierungseigenschaft steht in Zusammenhang mit einer Positivitätsbedingung für eine zyklische Matrix, deren Eigenwerte abgeschätzt werden. Diese lassen sich explizit durch die Matrixkoeffizienten angeben.

Aus der lokalen Minimierungseigenschaft für die nichtlokale Energie folgern wir letztendlich das Hauptresultat, nämlich die Existenz stark oszillierender, fast-periodischer lokaler Minimierer der ursprünglichen Funktionale.

MSC: 49J45, 49K40, 74N15, 74G65

Contents

Chapter 1

Introduction

Solid-to-solid phase transitions lead to fine mixtures of distinct phases with characteristic geometric structures. A phenomenon that occurs in martensitic phase transitions is that of twinning. Hereby, in a crystal near a critical temperature, a homogeneous phase called austenite, is separated from a martensite phase consisting of thin lamellae called "twins". The separation occurs at an interface between these two phases of the material (cf. [8], [29]). Mathematical models for the formation of such microstructures are based on the minimization of elastic energy, which was initiated by Khachaturyan [27], Khachaturyan and Shatalov [28] and Roitburd [44] in a linear model, while Ball and James ([8], [9]) introduced a nonlinear model based on this concept. To establish models that also give information on the characteristic scales such as the length of the twins, one has to include surface energy (see e.g. [8]), which has also important mathematical properties discussed later.

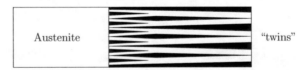

Figure 1.1: Austenite / twinned-martensite interface (cf. [29])

The role of nonlocal terms has become more and more important in the study of mathematical models for phase transitions. Singular perturbed functionals involving the $H^{1/2}$-norm have been considered by Alberti, Bouchitté and Seppecher in [5]. A nonlocal anisotropic model has been studied by Alberti and Bellettini (cf. [4], [3]). It is obtained from the Cahn-Hilliard model [11] by replacing the gradient term with the average of finite differences and has also been treated by Chmaj and Ren [13], [12]. A nonlocal model for the phase separation in diblock copolymer introduced by Ohta and Kawasaki [40] has been studied by Nishiura and Ohnishi [39] and by Ren and Wei in [41] and [43] in the one-dimensional case and in [42] for a two-dimensional model. Another example of nonlocal variational problems are models for micromagnetic materials (e.g. [20]).

Following Kohn and Müller [29], we rewrite a two-dimensional model for an austenite / twinned-martensite phase transformation to derive a nonlocal one-dimensional problem that describes the situation along the interface. The minimization problem they have considered is given by

$$\min_u \left[\int_{-1}^{1} \int_{0}^{L} (\varepsilon^2 u_{yy}^2 + W(u_y) + \alpha u_x^2) dx dy + \int_{-1}^{1} \int_{-\infty}^{0} (\beta_1 u_x^2 + \beta_2 u_y^2) dx dy \right]. \qquad (1.1)$$

Here, $u : (-\infty, L) \times (-1, 1) \to \mathbb{R}$ is any function periodic in y, i.e. $u(x, -1) = u(x, 1)$ for every $x \in (-\infty, L)$. It describes the shape of the material. $W : \mathbb{R} \to \mathbb{R}$ is a symmetric double-well potential with zeros -1, $+1$ only and $W(v) > 0$ elsewhere (see Figure 1.2). ε, α, β_1 and β_2 are material constants.

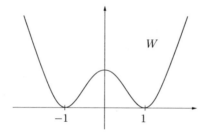

Figure 1.2: two-well potential

The domain $(0, L) \times (-1, 1)$ models the martensite, and

$$\int_{-1}^{1} \int_{0}^{L} (\alpha u_x^2 + W(u_y)) dx dy$$

is the elastic energy in the martensite. The strain-gradient term

$$\int_{-1}^{1} \int_{0}^{L} \varepsilon^2 u_{yy}^2 dx dy$$

represents the surface energy. $(-\infty, 0) \times (-1, 1)$ models the austenite, and

$$\int_{-1}^{1} \int_{-\infty}^{0} (\beta_1 u_x^2 + \beta_2 u_y^2) dx dy \qquad (1.2)$$

is the energy in the austenite. A mathematical treatment of this model can be found in [30]. As in [29], we "restrict" the functional to the interface $\{0\} \times (-1, 1)$: Since the two phases only share this boundary, minimization can be done over each of the domains separately for any given boundary $u_0(y) = u(0, y)$.

As for the austenite phase, the Euler-Lagrange equation is given by

$$\beta_1 u_{xx} + \beta_2 u_{yy} = 0,$$

and the unique solution u of this equation under the condition $u(0, y) = u_0(y)$ is given by

$$u(x, y) = \sum_{k=-\infty}^{\infty} \hat{u}_0(k) \exp(ik\pi y) \exp\left[\left(\frac{\beta_2}{\beta_1}\right)^{1/2} k\pi x\right]$$

provided that

$$u_0(y) = \sum_{k=-\infty}^{\infty} \hat{u}_0(k) \exp(ik\pi y)$$

is the Fourier expansion of u_0. Setting u into (1.2), we obtain

$$\int_{-1}^{1} \int_{-\infty}^{0} (\beta_1 u_x^2 + \beta_2 u_y^2) dx dy = 2\pi(\beta_1\beta_2)^{1/2} \sum_{k=-\infty}^{\infty} |k||\hat{u}_0(k)|^2.$$

Thus, a one-dimensional model for the interface between austenite and martensite involving the influence of the (two-dimensional) energy from the austenite is given by

$$I^\varepsilon(u) = \int_{-1}^{1} \varepsilon^2(u'')^2 + W(u')dy + 2\pi\sigma \sum_{k=-\infty}^{\infty} |k||u_k|^2, \tag{1.3}$$

where u is any periodic function (i.e. $u(-1) = u(1)$), and u_k, $k \in \mathbb{Z}$, are the Fourier coefficients, i.e.

$$u_k = \frac{1}{2} \int_{-1}^{1} u(x) \exp(-ik\pi x) dx, \quad u(x) = \sum_{k=-\infty}^{\infty} u_k \exp(ik\pi x).$$

It has also been proposed in [6], Section 6.4. Since (1.3) is invariant under the addition of constants, we may restrict to functions u with $\int_{-1}^{1} u = 0$. The term involving Fourier coefficients is the squared $H^{1/2}$-norm, which can be written in the form (cf. [6], Section 6.4)

$$\|u\|_{H^{1/2}}^2 = 2\pi \sum_{k=-\infty}^{\infty} |k||u_k|^2 = \int_{-1}^{1} \int_{-1}^{1} g(x - y)(u(x) - u(y))^2 dx dy =: E(u), \tag{1.4}$$

where

$$g(t) = \frac{\pi}{4(1 - \cos(\pi t))}. \tag{1.5}$$

The $H^{1/2}$-norm can also be obtained by interpolation between $H^1(-1, 1)$ and $L^2(-1, 1)$ with parameter $\theta = \frac{1}{2}$ (see [32], §9). The most interesting (and challenging) property of $\|u\|_{H^{1/2}}^2$ is that it is nonlocal, which in [3] has been characterized as follows: Given two disjoint intervals A and B, we have $E(u, A \cup B) > E(u, A) + E(u, B)$. Hereby, the error is the nonlocal interaction which due to the symmetry of g is given by

$$2 \int_A \int_B g(x - y)(u(x) - u(y))^2 dx dy.$$

In other words, nonlocality means that modifications of a function u on an interval A have effects on the energy via the whole interval $(-1, 1)$ and not only through the local energy on A.

The asymptotic behavior of minimizers of functionals similar to (1.3) for $\varepsilon \to 0$ have been studied extensively in literature, although with a right-hand term different from and easier to handle than the $H^{1/2}$-norm. It has been shown in [36] and [37] that the minimizers u_ε of the functionals

$$J^\varepsilon(u) = \int_{-1}^{1} \varepsilon^2 (u'')^2 + W(u') + u^2 dx \qquad (1.6)$$

show an increasingly oscillating behavior as $\varepsilon \to 0$, and that intervals on which $u_\varepsilon' \approx -1$ or $u_\varepsilon' \approx 1$, respectively, are separated by transition layers of size $\sim \varepsilon$ on which u_ε' jumps between -1 and 1. Indeed, the inclusion of surface energy in form of a second-order term penalizing these transitions is also of mathematical importance, since it provides lower semi-continuity of J^ε for $\varepsilon > 0$, which implies existence of minimizers. This does not hold if the singular perturbation is omitted, since any sequence $(u_n)_{n \in \mathbb{N}}$ converging to zero and satisfying $u_n'(x) \in \{-1, 1\}$ for a.e. x (with a finite number of jumps) is a minimizing sequence for J^0, which is due to $J^0(u) \geq 0$ and to the fact that -1 and 1 are the only zeros of W. The second-order term delivers a criterion to select particular minimizing sequences which have a "regular" structure. The main result in [37] is

Theorem 1.1 (Müller 1993) *There exists $\varepsilon_0 > 0$ such that for every $\varepsilon \in (0, \varepsilon_0)$, the minimizers u_ε of J^ε in $H^2(-1, 1)$ under periodic boundary conditions (i.e. $u(-1) = u(1)$, $u'(-1) = u'(1)$) or homogeneous Dirichlet boundary conditions (i.e. $u(-1) = u(1) = 0$) are periodic with minimal period*

$$P^\varepsilon \sim \varepsilon^{1/3},$$

and minimal energy

$$J^\varepsilon(u_\varepsilon) \sim \varepsilon^{2/3}.$$

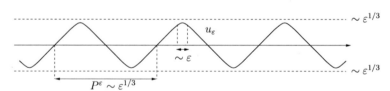

Figure 1.3: Minimizer u_ε of J^ε

A new method presented by Alberti and Müller in [6] considers suitably rescaled functions rather than the functions themselves. It has been applied to functionals more general than (1.6) in the sense that a positive weight function a is included in the L^2-term. Since Theorem 1.1 suggests that $\varepsilon^{1/3}$ is the microscopic scale, Alberti and Müller have rewritten the functionals in terms of functions $R^\varepsilon u : (-1, 1) \to K$, the "$\varepsilon$-blow-ups", K being a suitable compact function space. These are defined by

$$x \mapsto R_x^\varepsilon u \in K$$

$$R_x^\varepsilon u(t) = \varepsilon^{-1/3} u(x + \varepsilon^{1/3} t) \tag{1.7}$$

and describe the local "pattern" of u near any $x \in (-1, 1)$. Provided that u_ε is a minimizer for every $\varepsilon > 0$, they have shown that the sequence $R^\varepsilon u_\varepsilon$ generates a Young measure ν, and for every x, ν_x is supported on the set of all translations of a "regular" sawtooth function v whose corners are equidistant and whose period depends on $a(x)$, a being the weight function.

As proposed in Section 6.4 of [6], this approach can also be applied to the $H^{1/2}$-problem (1.3) for $\sigma = 1$, although the correct scale is given by $\varepsilon^{1/2}$ rather than $\varepsilon^{1/3}$. The conjecture formulated therein has indeed been the motivation for this work: In the case where $\sigma = 1$, Alberti and Müller have assumed that the ε-blow-ups $R^\varepsilon u_\varepsilon$ of the minimizers u_ε of (1.3) generate a Young measure ν with $\nu_x = \mu$ for a.e. x, where the probability measure μ on K is supported on the set of translations of a sawtooth function with equidistant corners. This would imply that for small $\varepsilon > 0$, the minimizers u_ε of I^ε in $H^2(-1, 1)$ under periodic boundary conditions are close to periodic functions with period $P^\varepsilon \sim \varepsilon^{1/2}$ and $u' = \pm 1$ everywhere and look like the ones of J^ε shown in Figure 1.3 with $\varepsilon^{1/3}$ replaced by $\varepsilon^{1/2}$.

To minimize the limit functional, however, turned out to be a very difficult problem due to the nonlocal structure of the $H^{1/2}$-norm. In this work, we examine under which conditions the conjectured global minimizers of I^ε are local minimizers of a similar, but quite general nonlocal variational problem. An essential difference to the models cited above is the use of a different parameter range, $\sigma \sim \varepsilon$, which has the advantage that the limit functional can be derived more easily. This parameter range has also been considered by Ren and Wei in [41] for functionals similar to (1.3). There, the one-dimensional case of a model for the micro-phase separation in diblock-copolymer introduced by Ohta and Kawasaki (see [39] and references therein) is treated mathematically. A special case of these functionals is given by

$$F^\varepsilon(v) = \int_0^1 \varepsilon(v')^2 + \frac{1}{\varepsilon} W(v) + |(-\Delta)^{-1/2} v|^2 dx, \quad v \in H^1(0,1) \cap \left\{ \int_0^1 v(x) dx = 0 \right\}, \tag{1.8}$$

where $(-\Delta)$ is the Laplacian operator given by

$$(-\Delta) : \left\{ u \in H^2(0,1) \;\middle|\; \int_0^1 u = 0,\; u'(0) = u'(1) = 0 \right\} \to \left\{ v \in L^2(0,1) \;\middle|\; \int_0^1 v = 0 \right\}$$

viewed as an operator on $L^2(0,1) \cap \{ \int u = 0 \}$. Then $(-\Delta)$ is an isomorphism, and $(-\Delta)^{-1}$ is positive and self-adjoint. $(-\Delta)^{-1/2}$ is its square root, which is a nonlocal operator. The following property is a special case of the main theorem in [41] and states that for small ε, there are local minimizers which look more or less like functions taking values -1 and $+1$ only, the jumps being equidistant.

Theorem 1.2 (Ren/Wei 2000) *For every $N \in \mathbb{N}$, there exists $\varepsilon_0 > 0$ such that for every $\varepsilon \in (0, \varepsilon_0)$, we find an $L^2(0,1)$-local minimizer v_ε of F^ε with*

$$\lim_{\varepsilon \to 0} \| v_\varepsilon - \bar{v}_N \|_{L^2(0,1)} = 0$$

where $\bar{v}_N$ is given by

$$\bar{v}_N(x) = (-1)^{i-1} \quad for \quad x \in (x_{i-1}, x_i), \quad i = 1, \ldots, N+1$$

with

$$x_0 = 0, \quad x_1 = \frac{1}{2N}, \quad x_i = x_{i-1} + \frac{1}{N} \quad for \quad i = 2, \ldots, N, \quad x_{N+1} = 1.$$

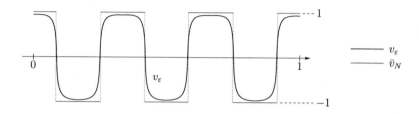

Figure 1.4: Local minimizer v_ε of F^ε

Setting $v = u'$, where $u(0) = 0$ and hence $u(0) = u(1) = 0$, a simple calculation which makes use of the self-adjointness of $(-\Delta)^{-1/2}$ yields

$$F^\varepsilon(v) = \int_0^1 \varepsilon(u'')^2 + \frac{1}{\varepsilon}W(u') + u^2 dx,$$

which essentially is (1.6) with a different parameter range. In particular, rewriting (1.8) in terms of antiderivatives u of v turns the nonlocal term in u into a local one in v. Or vice versa, rewriting (1.6) in terms of derivatives of v yields (1.8), from which one can derive a natural generalization of (1.6) to higher dimensions.

The approach in [41] is a different one than in [37] or [6] and makes use of the particular structure of the nonlocal term. Most of the methods used therein, however, are not transferrable to a general nonlocal model.

We return to (1.3), which by (1.4) has the form

$$I^\varepsilon(u) = \int_{-1}^1 \varepsilon^2(u'')^2 + W(u')dy + \sigma E(u),$$

where the nonlocal energy is

$$E(u) = \int_{-1}^1 \int_{-1}^1 g(x-y)(u(x)-u(y))^2 dx dy,$$

and the space of admissible functions is the set of all $u \in H^2(-1,1)$ with $u(-1) = u(1)$, $u'(-1) = u'(1)$ and $\int u = 0$.

As I have mentioned, we use the same parameter range $\sigma = \varepsilon$ as in [41], which mathematically has the advantage that the Modica-Mortola theorem can be applied "immediately" to obtain a limit functional I (i.e. the Γ-limit of $(I^\varepsilon)_{\varepsilon>0}$) without having to consider rescaled functions. Indeed, the rescaling (1.7) proposed by Alberti and Müller in [6] for the functionals (1.6) is necessary to obtain a well-defined limit problem via the Modica-Mortola theorem.

Instead of the unbounded kernel (1.5), we consider a bounded, but general kernel g that has the same symmetry and periodicity properties as (1.5) (i.e. g, extended to $\mathbb{R}$ periodically, is 2-periodic and symmetric).

Thus, the model we consider is an approximation for (1.3) (with $\sigma = \varepsilon$), and in the special case where $g = 1$, we obtain $E(u) = 4\|u\|^2_{L^2(-1,1)}$ due to the zero average condition, so that I^ε in a way is a generalization of (1.8). We aim to show an analog to Theorem 1.2, i.e. for small $\varepsilon > 0$, we construct local minimizers u^ε of I^ε whose derivatives are close to functions that switch between -1 and $+1$ and whose jumps lie equidistantly in $(-1, 1)$.

The framework we use is similar to that used in [41], and some properties and methods mentioned therein are applied in this work in a modified form. In the proof of the main result formulated in the following chapter, however, the quite general nonlocal term makes it necessary to apply a lot of particular techniques. The outline of this work is as follows:

In Chapter 2, we formulate the problem in detail and give some simple properties as well as presenting basic techniques that are used throughout the work. In Chapter 3, we prove a version of the Modica-Mortola theorem, which leads to a limit functional I derived in Chapter 4, where we will also show how local minimizers of I^ε can be obtained from those of I.

The limit functional I is only finite on the set of periodic sawtooth functions u with $u' \in \{-1, 1\}$ on $(-1, 1)$, a finite number of corners and zero average. Since we are interested in local minimizers of I, we consider admissible variations on the set of sawtooth functions in Chapter 5, and in chapter 6 we transfer the results to the case of sawtooth functions with equidistant corners, which we conjecture to be local minimizers of I. In chapters 7 and 8, we show how the local minimality property of these particular functions is connected to the positivity of eigenvalues of certain matrices $A_N \in \mathbb{R}^{N \times N}$, N denoting a fixed number of corners, while chapters 9 and 10 deal with the positivity of these eigenvalues.

Acknowledgements: I would like to thank Prof. Hansjörg Kielhöfer for giving me the opportunity to write this thesis and for always giving me good advice during the years I have been working on it. I want to express my gratitude to Prof. Stefan Müller for inviting me to the Max-Planck-Institute for Mathematics in the Sciences in Leipzig twice, for pointing me to this fascinating subject and for numerous very stimulating and fruitful discussions. I want to thank Prof. Timothy Healey for his invitation to the Department of Theoretical and Applied Mechanics at Cornell University, where I had a great opportunity to present first results and to exchange ideas. My gratitude also goes to Prof. Stanislaus Maier-Paape for inviting me to the RWTH Aachen and for inspiring discussions. Finally, I would like to thank the Graduiertenkolleg "Nichtlineare Probleme in Analysis, Geometrie und Physik" at the University of Augsburg for financial support between 1999 and 2001.

Chapter 2

Preliminaries

2.1 The Functionals I^ε

We consider the functionals

$$I^\varepsilon(u) = \int_{-1}^{1} \varepsilon(u'')^2 + \frac{1}{\varepsilon} W(u') dx + E(u), \quad u \in H^{2,0}_{per}(-1, 1), \tag{2.1}$$

for $\varepsilon > 0$, where

$$H^{2,0}_{per}(-1, 1) = \{u \in H^2(-1, 1) \mid u(-1) = u(1), \ u'(-1) = u'(-1), \ \int_{-1}^{1} u(x) dx = 0\}$$

is the Hilbert space of all periodic H^2-functions $u : (-1, 1) \to \mathbb{R}$ with zero average, endowed with the usual norm $\|\cdot\|_{H^2(-1,1)}$. $W : \mathbb{R} \to \mathbb{R}$ is a double-well potential like in Figure 1.2 satisfying certain conditions specified later on.

E is the nonlocal energy defined by

$$E(u) = \int_{-1}^{1} \int_{-1}^{1} g(x - y)(u(x) - u(y))^2 dx dy,$$

where the kernel $g : \mathbb{R} \to \mathbb{R}$ satisfies the following conditions:

(g_1) $g(x + 2) = g(x)$ for every $x \in \mathbb{R}$

(g_2) $g(x) = g(-x)$ for every $x \in \mathbb{R}$

(g_3) There exist positive constants g_0, g_1 with $g_0 \leq g(x) \leq g_1$ for every $x \in \mathbb{R}$

(g_4) $g\big|_{[0,1]} \in C^3[0, 1]$

(g_5) $g'(1) = 0$

(g_6) g is globally Lipschitz continuous on $\mathbb{R}$.

Remarks:

a) (g_1) and (g_2) imply $g(1-x) = g(1+x)$, i.e. g is 2-periodic and symmetric with respect to $x = 1$, $x = 0$. Thus, g is symmetric with respect to every $k \in \mathbb{Z}$.

b) Condition (g_6) is redundant: Due to the symmetry of g with respect to every $k \in \mathbb{Z}$, (g_4) together with (g_3) implies $g \in W^{1,\infty}(\mathbb{R})$. Thus, for every $x, y \in \mathbb{R}$, $y > x$, we get

$$|g(x) - g(y)| = \left| \int_x^y g'(\xi)d\xi \right| \leq C|x - y|,$$

which shows the Lipschitz continuity.

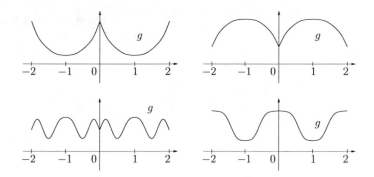

Figure 2.1: Examples for the kernel g

The double-well potential $W : \mathbb{R} \to \mathbb{R}$ has to satisfy the following conditions:

(W_1) $W(v) \geq 0$ for every $v \in \mathbb{R}$, and $W(v) = 0 \Leftrightarrow v \in \{-1, 1\}$

(W_2) $W(-v) = W(v)$ for every $v \in \mathbb{R}$

(W_3) There are $c > 0$, $u_0 > 0$ such that $W(v) \geq cv^2$ for every $v \in \mathbb{R}$ with $|v| \geq u_0$

(W_4) $W \in C^2(\mathbb{R})$

(W_5) $W'(-1) = W'(1) = 0$, $W''(-1) > 0$, $W''(1) > 0$

Remarks:

a) The most common example used in many works (e.g. [36], [31]) is $W(v) = (v^2 - 1)^2$.

b) The symmetry (W_2) is only required to simplify the proof of the upper bound condition in the version of the Modica-Mortola theorem we show in Chapter 3, and all results we show will also hold if this condition is dropped.

c) The growth condition at infinity (W_3) is needed in the proof of the Modica-Mortola theorem. To derive the Γ-limit of the respective functionals for $\varepsilon \to 0$ with respect to the L^1-topology as in the classical works by Modica [33] and Modica / Mortola [34], [35] (also see [2]), at least linear growth is required. For the version proved in this work, which uses the L^2-topology as in Ren/Wei [41], quadratic growth is needed.

d) Since W is C^2 and has the global (and thus local) minimum 0 at $v = -1$ and $v = 1$, the conditions $W'(-1) = W'(1) = 0$ are redundant.

We also have the following property, which is used in the next chapter:

Lemma 2.1 *If W satisfies (W_1)-(W_5), then $v \mapsto \sqrt{W(v)}$ is locally Lipschitz continuous.*

Proof: Since -1, 1 are the only zeros of W on $\mathbb{R}$, $v \mapsto \sqrt{W(v)}$ is C^1 and thus locally Lipschitz continuous on $\mathbb{R} \setminus \{-1, 1\}$, so that we only have to show the local Lipschitz condition near -1 and 1. We restrict to the latter, the former is shown in an analogous way. Since $W''(1) > 0$ and W is C^2, we can choose $\delta > 0$ such that $W''(\xi) > 0$ for every $\xi \in [1 - \delta, 1 + \delta]$. Let $C_\delta > 0$ with $W''(\xi) \leq 2C_\delta$ for every $\xi \in [1 - \delta, 1 + \delta]$. Then for every $v \in [1 - \delta, 1 + \delta]$, we find $\xi \in [1 - \delta, 1 + \delta]$ such that

$$W(v) = W(1) + W'(1)(v - 1) + \frac{1}{2}W''(\xi)(v - 1)^2 = \frac{1}{2}W''(\xi)(v - 1)^2 \leq C_\delta(v - 1)^2,$$

where we have used $W(1) = 0$, $W'(1) = 0$. We conclude

$$|\sqrt{W(v)} - \sqrt{W(1)}| = \sqrt{W(v)} \leq \sqrt{C_\delta(v - 1)^2} = \sqrt{C_\delta}|v - 1|,$$

which shows the local Lipschitz continuity. $\square$

2.2   Existence of Minimizers

Before we formulate the main theorem, we will prove that for every $\varepsilon > 0$, I^ε has a minimizer. Roughly speaking, this holds since I^ε is convex in the highest derivative (see [14], Chapter 3, Theorem 4.1). First, we show the following equivalence of E and $\| \cdot \|^2_{L^2(-1,1)}$ for functions with zero average:

Lemma 2.2 *For every $u \in L^2(-1, 1)$ with $\int_{-1}^{1} u(x)dx = 0$, we have*

$$4g_0\|u\|^2_{L^2(-1,1)} \leq E(u) \leq 4g_1\|u\|^2_{L^2(-1,1)},$$

where $\| \cdot \|_{L^2(-1,1)}$ denotes the usual norm on $L^2(-1, 1)$ and g_0, g_1 are the constants in (g_3). In particular, the inequalities hold for every $u \in H^{2,0}_{per}(-1, 1)$.

Proof: Let $u \in L^2(-1,1)$ with $\int_{-1}^{1} u(x)dx = 0$. Due to $g_0 \leq g \leq g_1$, we have

$$g_0 \int_{-1}^{1} \int_{-1}^{1} (u(x) - u(y))^2 dx dy \leq E(u) \leq g_1 \int_{-1}^{1} \int_{-1}^{1} (u(x) - u(y))^2 dx dy,$$

where

$$\int_{-1}^{1} \int_{-1}^{1} (u(x) - u(y))^2 dx dy = \int_{-1}^{1} \int_{-1}^{1} (u(x)^2 - 2u(x)u(y) + u(y)^2) dx dy$$

$$= 4 \int_{-1}^{1} u(x)^2 dx - 2 \left[\int_{-1}^{1} u(x) dx \right]^2 = 4\|u\|_{L^2(-1,1)}^2,$$

and the Lemma follows. $\qquad\square$

We are now able to prove the existence of minimizers of I^ε for $\varepsilon > 0$. Fix such an ε. Since $I^\varepsilon(u) \geq 0$ for every $u \in H_{per}^{2,0}(-1,1)$, we find a sequence $(u_n)_{n \in \mathbb{N}} \subset H_{per}^{2,0}(-1,1)$ with

$$\lim_{n \to \infty} I^\varepsilon(u_n) = \inf_{u \in H_{per}^{2,0}(-1,1)} I^\varepsilon(u). \tag{2.2}$$

In particular, we find $C_\varepsilon > 0$ with $I^\varepsilon(u_n) \leq C_\varepsilon$ for every $n \in \mathbb{N}$, so that, using Lemma 2.2,

$$\varepsilon \|u_n''\|_{L^2(-1,1)}^2 \leq I^\varepsilon(u_n) \leq C_\varepsilon, \quad 4g_0\|u_n\|_{L^2(-1,1)}^2 \leq E(u) \leq I^\varepsilon(u_n) \leq C_\varepsilon.$$

Since $\int_{-1}^{1} u_n'(x)dx = u(1) - u(-1) = 0$ and $u_n' \in H^1(-1,1) \hookrightarrow C[-1,1]$ (see [1]), we can choose $x_n \in [-1,1]$ such that $u_n'(x_n) = 0$ (where u_n' denotes the continuous representant of u_n'). Thus we get, using the Cauchy-Schwarz inequality,

$$|u_n'(x)| = \left| \int_{x_n}^{x} u_n''(\xi)d\xi \right| \leq \int_{-1}^{1} |u_n''(\xi)|d\xi \leq \sqrt{2}\|u_n''\|_{L^2(-1,1)} \leq \sqrt{2C_\varepsilon \varepsilon^{-1}}$$

for every $x \in [-1,1]$, so that

$$\|u_n'\|_{L^2(-1,1)}^2 = \int_{-1}^{1} (u_n'(x))^2 dx \leq 4C_\varepsilon \varepsilon^{-1},$$

hence $(u_n)_{n \in \mathbb{N}}$ is bounded in $H_{per}^{2,0}(-1,1)$, and we can choose a subsequence (not relabelled) such that

$$u_n \underset{n \to \infty}{\longrightarrow} \bar{u} \quad \text{in} \quad H_{per}^{2,0}(-1,1) \tag{2.3}$$

for a $\bar{u} \in H_{per}^{2,0}(-1,1)$. Since the imbedding $H_{per}^{2,0}(-1,1) \hookrightarrow C^1[-1,1]$ is compact (see [1]), we have $u_n \to \bar{u}$, $u_n' \to \bar{u}'$ uniformly, and we immediately deduce

$$\lim_{n \to \infty} \left[\int_{-1}^{1} \frac{1}{\varepsilon} W(u_n')dx + E(u_n) \right] = \int_{-1}^{1} \frac{1}{\varepsilon} W(\bar{u}')dx + E(\bar{u}). \tag{2.4}$$

The functional $u \mapsto \int_{-1}^{1} \varepsilon(u'')^2 dx$ clearly is convex and continuous on $H_{per}^{2,0}(-1,1)$, so that it is weakly lower semi-continuous (see [14], Chapter 3, Theorem 1.2), which due to (2.3) implies

$$\int_{-1}^{1} \varepsilon(\bar{u}'')^2 dx \leq \liminf_{n \to \infty} \int_{-1}^{1} \varepsilon(u_n'')^2 dx.$$

Thus, (2.4) and the definition of I^ε (2.1) yield

$$I^\varepsilon(\bar{u}) \leq \liminf_{n \to \infty} I^\varepsilon(u_n).$$

On the other hand, we derive from (2.2)

$$I^\varepsilon(\bar{u}) \geq \inf_{u \in H^{2,0}_{per}(-1,1)} I^\varepsilon(u) = \limsup_{n \to \infty} I^\varepsilon(u_n),$$

so that

$$I^\varepsilon(\bar{u}) = \lim_{n \to \infty} I^\varepsilon(u_n) = \inf_{u \in H^{2,0}_{per}(-1,1)} I^\varepsilon(u),$$

i.e. $\bar{u} \in H^{2,0}_{per}(-1,1)$ is a minimizer of I^ε.

2.3   The Main Result

As mentioned in the introduction, we aim to show that for small ε, I^ε has highly oscillating local minimizers in $H^{2,0}_{per}(-1,1)$ with a certain "regular" structure resembling sawtooth functions with slope ± 1 only and equidistant corners (see Figure 2.2). The main result in this work is the following:

Theorem 2.3 *For $\varepsilon > 0$, let I^ε as defined by (2.1). Then there exists $N_0 \in \mathbb{N}$ such that for every $N \geq N_0$ which is a multiple of 4, we find $\varepsilon_0 > 0$ and $\delta > 0$ satisfying the following condition:*

For every $\varepsilon \in (0, \varepsilon_0)$, there is a local minimizer $u_\varepsilon \in H^{2,0}_{per}(-1,1)$ of I^ε with respect to the $H^1(-1,1)$-norm such that

$$I^\varepsilon(u_\varepsilon) \leq I^\varepsilon(u) \quad \text{for every} \quad u \in H^{2,0}_{per}(-1,1) \quad \text{with} \quad \|u - \bar{u}_N\|_{H^1(-1,1)} < \delta$$

and

$$\lim_{\varepsilon \to 0} \|u_\varepsilon - \bar{u}_N\|_{H^1(-1,1)} = 0,$$

where $\bar{u}_N$ is a sawtooth function with N equidistant corners. More precisely, $\bar{u}_N$ is the unique function in $H^1(-1,1)$ satisfying $\int_{-1}^{1} \bar{u}_N(x)dx = 0$ and

$$\bar{u}'_N(x) = (-1)^{i-1} \quad \text{for} \quad x \in \left(-1 + \frac{2}{N}(i-1), -1 + \frac{2}{N}i\right), \quad i = 1, \ldots, N.$$

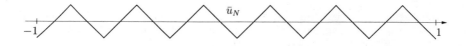

Figure 2.2: The function $\bar{u}_N$ for $N = 12$

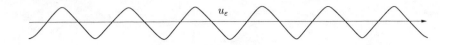

Figure 2.3: Near-periodic local minimizer u_ε resembling $\bar{u}_N$

Remarks:

a) The theorem states that if $\varepsilon > 0$ is chosen sufficiently small, the functionals I^ε have strongly oscillating local minimizers which look like a "regular" sawtooth function (as shown in Figure 2.2 for $N = 12$). Near the corners of $\bar{u}_N$, the local minimizers u_ε have transition layers where u'_ε switches between -1 and 1.

b) Since we consider periodic functions, only the case where N is even makes sense.

c) The result also holds if the condition that N is a multiple of 4 is replaced by the assertion that N is even. However, we only present the proof for the former case in full detail, while we only sketch the proof of the case where $\frac{N}{2}$ is odd (see section 10.5).

d) The result also holds if we replace the $H^1(-1,1)$-topology by the $W^{1,1}(-1,1)$-topology, and only minor modifications are required in the proof. The local minimizing property would then become a stronger one, while the convergence is weaker.

2.4 Γ-Convergence

To identify (local) minimizers of the functionals I^ε, we use De Giorgi's concept of Γ-convergence, which has been introduced in [18]. For a very general and comprehensive treatment of this theory, we refer the reader to [15].

For a metric space X, the Γ-limit of a sequence $(F^\varepsilon)_{\varepsilon>0}$ of functionals $F^\varepsilon : X \to [0,\infty]$ (if existent) is a limit functional $F : X \to [0,\infty]$ which - unlike pointwise convergence - makes it possible to characterize (local) minimizers of the functionals F^ε for small $\varepsilon > 0$ by considering (local) minimizers of F. The general definition of Γ-convergence can be found in [15], Definition 4.1. In metric spaces (or, more generally, topological spaces satisfying the first axiom of countability), Γ-convergence can be defined as follows (cf. [15], Proposition 8.1):

Definition 2.4 (Γ-convergence) *Let X be a metric space, and $(F^\varepsilon)_{\varepsilon>0}$ a family of functionals $F^\varepsilon : X \to [0,\infty]$. The functionals F^ε Γ-**converge** to the functional $F : X \to [0,\infty]$ for $\varepsilon \to 0$, denoted*

$$F^\varepsilon \xrightarrow{\;\Gamma\;} F,$$

if the following two conditions hold:

(i) For every $x \in X$ and for every sequence $(x_\varepsilon)_{\varepsilon>0} \subset X$ with $\lim_{\varepsilon \to 0} x_\varepsilon = x$ in X, we have

$$F(x) \le \liminf_{\varepsilon \to 0} F^\varepsilon(x_\varepsilon).$$

(ii) For every $x \in X$, there exists a sequence $(x_\varepsilon)_{\varepsilon>0} \subset X$ with $\lim_{\varepsilon \to 0} x_\varepsilon = x$ in X such that

$$\limsup_{\varepsilon \to 0} F^\varepsilon(x_\varepsilon) \le F(x).$$

To characterize Γ-convergence, we cite the following properties proved in [15] (see Proposition 6.8, Proposition 6.21 and Corollary 7.20 therein).

Proposition 2.5 *Let X be a metric space, $(F^\varepsilon)_{\varepsilon>0}$ be a family of functionals $F^\varepsilon : X \to [0, \infty]$, and $F : X \to [0, \infty]$ with $F^\varepsilon \xrightarrow{\Gamma} F$. Then*

(i) *F is lower semi-continuous on X*

(ii) *If $G : X \to [0, \infty)$ is continuous, then $F^\varepsilon + G \xrightarrow{\Gamma} F + G$.*

(iii) *For every $\varepsilon > 0$, let $u_\varepsilon \in X$ be a minimizer of F^ε. Assume that there exists $u \in X$ with $\lim_{\varepsilon \to 0} u_\varepsilon = u$ in X. Then u is a minimizer of F, and*

$$\lim_{\varepsilon \to 0} F^\varepsilon(u_\varepsilon) = F(u).$$

Assertion (iii) can be described as an "invariance" of the minimizing property for the functionals F^ε under the Γ-limit, so that minimizers of the latter allow us to characterize minimizers of F^ε for small ε. There are similar results on the connection between local minimizers on the Γ-limit F and those of F^ε for small ε (see Theorems 2.1 and 4.1 in [31] and §3 in [17]), which will be discussed later on.

Property (ii) and the Modica-Mortola theorem (see Chapter 3) will deliver the Γ-limit I of the functionals I^ε. Since we are looking for local minimizers of I^ε for small $\varepsilon > 0$, we will identify local minimizers of I and show how to obtain local minimizers of I^ε from the Γ-convergence property. Latter is done in Theorem 4.6 and Corollary 4.7.

2.5   Young Measures

In the proof of the Modica-Mortola theorem we give, which is based on the one by Alberti [2], we will apply the fundamental theorem for Young measures.

A Young measure ν associated with a sequence of functions $u_\varepsilon : \Omega \to K$ assigns a probability measure ν_s on K to every $s \in \Omega$ and thus describes the distribution of the values of the functions u_ε for $\varepsilon \to 0$. The concept originates from Young's idea of generalized solutions (see [47], Chapter VI) and has ever since developed to a crucial technique in the calculus of variations, and to a great extent in the examination of minimizing sequences of non-convex functionals. For further information on Young measures, see §3 in [38], or [45]. The approach described below is taken from [6], §2 (also see [7]).

Let $\Omega \subset \mathbb{R}^k$ open, $K \subset \mathbb{R}^n$ compact, λ^k the k-dimensional Lebesgue measure on Ω. Let $\mathcal{M}(K)$ be the Banach space of finite signed Borel measures on K with norm $\|\mu\| = \int_K d|\mu|$, which is the total variation of the measure μ (see [22], Kapitel VII, §1). Then the dual space of $C(K)$ can be identified with $\mathcal{M}(K)$ via the duality pairing

$$\langle \mu, g \rangle = \int_K g d\mu$$

(see Kapitel VIII, §2 in [22] or Section 4.10 in [21]), which also defines the weak*-topology on $\mathcal{M}(K)$. Then by [21], section 8.18, the dual space of $L^1(\Omega, C(K))$ can be identified with the space $L^\infty_w(\Omega, \mathcal{M}(K))$ of weak*-measurable maps $\nu : \Omega \to \mathcal{M}(K)$ via the duality

$$\langle \nu, \phi \rangle = \int_\Omega \langle \nu_s, \phi_s \rangle ds. \qquad (2.5)$$

Here, weak*-measurable means that the pre-image of every $M \subset \mathcal{M}(K)$ which lies in the Borel σ-algebra generated by the weak*-open sets in $\mathcal{M}(K)$ is an element of the Borel σ-algebra generated by the open sets in Ω. Obviously, if $\nu \in L^\infty_w(\Omega, \mathcal{M}(K))$ has the form

$$\nu_s = \alpha(s)\delta_a + (1 - \alpha(s))\delta_b,$$

with $\alpha : \Omega \to [0,1]$, where δ_y denotes the Dirac measure in $y \in K$, the function α is Borel measurable.

Definition 2.6 *Let* $\Omega \subset \mathbb{R}^k$ *open,* $K \subset \mathbb{R}^n$ *compact. Then* $\nu \in L^\infty_w(\Omega, \mathcal{M}(K))$ *is a* **Young measure** *on* Ω *with values in* K *if* ν_s *is a probability measure for a.e.* $s \in \Omega$. *The set of all Young measures* $\nu \in L^\infty_w(\Omega, \mathcal{M}(K))$ *is denoted by* $\mathcal{Y}(\Omega, K)$.

A compactness result for $\mathcal{Y}(\Omega, K)$ in $L^\infty_w(\Omega, \mathcal{M}(K))$ proved in [10] is the abstract foundation for the fundamental theorem for Young measures. It has been shown in [7] in a very general form, a special case of which is given by

Theorem 2.7 (Fundamental Theorem for Young measures) *Let* $\Omega \subset \mathbb{R}^k$ *be an open set,* $K \subset \mathbb{R}^m$ *compact. Let* $(u_n)_{n \in \mathbb{N}}$ *be a sequence of functions* $u_n : \Omega \to K$. *Then there exists a subsequence of* $(u_n)_{n \in \mathbb{N}}$ *(not relabelled) and a Young measure* $\nu \in \mathcal{Y}(\Omega, K)$ *such that*

$$\lim_{n \to \infty} \int_\Omega g(u_n(s))h(s)ds = \int_\Omega \int_K g(x)d\nu_s(x)ds$$

for every $g \in C(K)$, $h \in L^1(\Omega)$.

Remark: In view of the duality pairing (2.5), this means that $\nu^n \overset{*}{\rightharpoonup} \nu$, where ν^n is the elementary Young measure associated with u_n, defined by $\nu^n_s = \delta_{u_n(s)}$ for every $s \in \Omega$, $n \in \mathbb{N}$ (see §2 in [6]).

2.6   Remarks on Notation

We mention some important notation used throughout the work. For a set $I \subset \mathbb{R}^n$, χ_I denotes the characteristic function of I. For given numbers $a, b \in \mathbb{R}$, $a \vee b$ is the maximum of a and b, while $a \wedge b$ denotes the minimum of a and b. For $n \in \mathbb{N}$, λ^n denotes the Lebesgue measure on $\mathbb{R}^n$. Let $\Omega \subset \mathbb{R}^n$ be an open set, then $C_0^\infty(\Omega)$ is the set of all functions that are of class C^∞ on Ω and whose support is a compact set in Ω. For a complex number $z = a + ib$, $Re(z) = a$ is the real part of z, while $Im(z) = b$ is the imaginary part of z. We will also make use of generic constants: Constants like c, C etc. may grow in every step of an estimation, but are only dependent on the initial data (such as g_0, g_1, W) and independent on ε and N, unless otherwise stated.

Chapter 3

A Version of the Modica-Mortola Theorem

3.1 The Modica-Mortola Theorem

In this chapter, we formulate a version of the classical Modica-Mortola theorem, which in its original form is found in [33], [34], [35]. The main differences in this version are the use of the L^2-topology rather than the L^1-topology and that the functionals are defined for periodic functions, which leads to a slightly different counting term (i.e. 0-dimensional Hausdorff measure) in the limit functional. Furthermore, we always consider functions $u : (-1, 1) \to \mathbb{R}$ with an arbitrary range instead of $u : (-1, 1) \to [-1, 1]$, which makes it necessary to apply truncation (also see Lemma 1.14 in [3]). Although only minor modifications have to be done to obtain the result, a complete proof of the theorem will be given. The proof is essentially the one given by Alberti [2], which uses the fundamental theorem of Young measures. Throughout the chapter, let X be the Hilbert space

$$X = L^2(-1, 1) \cap \left\{ \int_{-1}^{1} v = 0 \right\}$$

endowed with norm $\| \cdot \| = \| \cdot \|_{L^2}$. Furthermore, define a constant A_0 depending on W only, by

$$A_0 = 2 \int_{-1}^{1} \sqrt{W(v)} dv. \tag{3.1}$$

Let

$$H_{per}^{1,0}(-1, 1) = \{ u \in H^1(-1, 1) \mid u(-1) = u(1), \int_{-1}^{1} u(x) dx = 0 \}$$

be the set of all periodic H^1-functions $u : (-1, 1) \to \mathbb{R}$ with zero average. We define a family of functionals $(J^\varepsilon)_{\varepsilon > 0}$, $J^\varepsilon : X \to [0, \infty]$ and a functional $J : X \to [0, \infty]$, given by

$$J^\varepsilon(v) = \begin{cases} \int_{-1}^{1} \varepsilon(v')^2 + \frac{1}{\varepsilon} W(v) dx & \text{for } v \in H_{per}^{1,0}(-1, 1), \\ \infty & \text{for } v \in X \setminus H_{per}^{1,0}(-1, 1) \end{cases} \tag{3.2}$$

17

$$J(v) = \begin{cases} A_0 \#(Sv \cap [-1,1)) & \text{for } v \in X \cap BV((-1,1), \{-1,1\}) \\ \infty & \text{for } v \in X \setminus BV((-1,1), \{-1,1\}). \end{cases} \qquad (3.3)$$

Here, $BV((-1,1), \{-1,1\})$ is the set of all functions $v : (-1,1) \to \{-1,1\}$ with bounded variation. This is exactly the set of functions $v : (-1,1) \to \mathbb{R}$ which only take values -1 and $+1$ and have a finite number of jumps, i.e. discontinuity points. For a general definition of BV functions, see [23]. For our purposes, however, this characterization will do.

$\#(Sv \cap [-1,1))$ denotes the number of jumps (i.e. discontinuity points) in the semi-open interval $[-1,1)$, which means that an additional jump will be added to the inner jumps if the 2-periodic extension of v to $\mathbb{R}$ is discontinuous at $t = -1$.

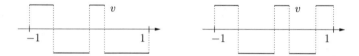

Figure 3.1: Two functions with $\#(Sv \cap [-1,1)) = 4$.

With the definitions above, we are now able to formulate the following version of the Modica-Mortola theorem:

Theorem 3.1 *(i) Let $(\varepsilon_n)_{n \in \mathbb{N}} \subset \mathbb{R}^+$ such that $\varepsilon_n \to 0$ and $(v_n)_{n \in \mathbb{N}} \subset H_{per}^{1,0}(-1,1)$ such that $(J^{\varepsilon_n}(v_n))_{n \in \mathbb{N}}$ is a bounded sequence in $\mathbb{R}$. Then the sequence $(v_n)_{n \in \mathbb{N}}$ is relatively compact in X, and every cluster point of $(v_n)_{n \in \mathbb{N}}$ with respect to the topology in X lies in $X \cap BV((-1,1), \{-1,1\})$.*

(ii) Let $(v_\varepsilon)_{\varepsilon > 0} \subset H_{per}^{1,0}(-1,1)$, $v \in X$ such that $\|v_\varepsilon - v\|_{L^2(-1,1)} \to 0$ for $\varepsilon \to 0$. Then

$$J(v) \leq \liminf_{\varepsilon \to 0} J^\varepsilon(v_\varepsilon).$$

(iii) For every $v \in X \cap BV((-1,1), \{-1,1\})$, there exists a sequence $(v_\varepsilon)_{\varepsilon > 0} \subset H_{per}^{1,0}(-1,1)$ such that $\|v_\varepsilon - v\|_{L^2(-1,1)} \to 0$ for $\varepsilon \to 0$ and

$$\limsup_{\varepsilon \to 0} J^\varepsilon(v_\varepsilon) \leq J(v).$$

In particular,

$$J^\varepsilon \xrightarrow{\Gamma} J \quad in \quad X.$$

3.2 Preparations

Before we prove Theorem 3.1, we introduce some useful tools which are taken from [2]. In particular, we will have to consider the integral term in (3.2) on a set of functions which is larger than $H_{per}^{1,0}(-1, 1)$, since modifications of functions in $H_{per}^{1,0}(-1, 1)$ needed in the proof may not be periodic or have zero average.

3.2.1 Localization

For a measurable set $A \subset \mathbb{R}$ and any function $v : A \to \mathbb{R}$ which has a distributional derivative $v' \in L^2(U)$, U some open set with $A \subset U$, we set

$$K^\varepsilon(v; A) = \int_A \varepsilon(v')^2 + \frac{1}{\varepsilon}W(v)dx.$$

3.2.2 Scaling Property

Let $v : A \to \mathbb{R}$ be as in 3.2.1. For $\varepsilon > 0$, let $\frac{1}{\varepsilon}A = \{t \in \mathbb{R} \,|\, \varepsilon t \in A\}$ and define a rescaled function $v^\varepsilon : \frac{1}{\varepsilon}A \to \mathbb{R}$ by $v^\varepsilon(t) = v(\varepsilon t)$. Then the following scaling property holds:

$$K^\varepsilon(v^\varepsilon; \frac{1}{\varepsilon}A) = K^1(v; A). \tag{3.4}$$

3.2.3 The Optimal Profile

We define the optimal profile as

$$\bar{\sigma} = \inf \left\{ K^1(v; \mathbb{R}) \,\middle|\, v : \mathbb{R} \to [-1, 1], \; v' \in L^2(\mathbb{R}), \; \lim_{s \to -\infty} v(s) = -1, \; \lim_{s \to +\infty} v(s) = 1 \right\}, \tag{3.5}$$

where v' denotes the distributional derivative of v.

3.2.4 Other Useful Properties

We will show some important features needed in the proof of Theorem 3.1.

Lemma 3.2 *Let $I \subset \mathbb{R}$ be an interval, $v : I \to [-1, 1]$ which has a distributional derivative $v' \in L^2(I)$. Let $a, b \in I$, $\delta > 0$ such that $v(a) \leq -1 + \delta$, $v(b) \geq 1 - \delta$. Then*

$$K^\varepsilon(v; I) \geq \bar{\sigma} - C_1\delta \quad \text{for all } \varepsilon > 0,$$

where C_1 is a constant independent of δ, ε, u and I.

Proof: We may restrict to the case $a < b$. Due to the monotonicity $K^\varepsilon(v; I) \geq K^\varepsilon(v; [a, b])$ and the scaling property (3.4), we may assume $I = [a, b]$ and $\varepsilon = 1$, respectively. Given $v(a) \leq -1 + \delta$ and $v(b) \geq 1 - \delta$, we extend v to the whole real line as follows:

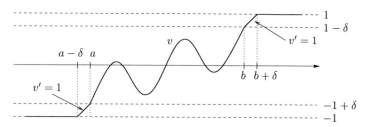

Figure 3.2: extension of v

Then $|v| \leq 1$ so that $W(v)$ is bounded, and since both v' and $W(v) = W(\pm 1)$ vanish outside of $(a - \delta, b + \delta)$, we obtain

$$K^1(v; \mathbb{R} \setminus I) = \int_{a-\delta}^{a} dx + \int_{b}^{b+\delta} dx + \int_{a-\delta}^{a} W(v) dx + \int_{b}^{b+\delta} W(v) dx \leq C_1 \delta.$$

Obviously, $v' \in L^2(\mathbb{R})$ with $\lim_{x \to \pm \infty} v(x) = \pm 1$, and the definition of the optimal profile (3.5) yields

$$K^1(v; I) = K^1(v, \mathbb{R}) - K^1(v; \mathbb{R} \setminus I) \geq \bar{\sigma} - C_1 \delta.$$

$\square$

Lemma 3.3 *The infimum $\bar{\sigma}$ given by (3.5) is a minimum, and its value is*

$$\bar{\sigma} = 2 \int_{-1}^{1} \sqrt{W(v)} dv = A_0,$$

A_0 given by (3.1). Furthermore, (3.5) has a minimizer γ which is strictly increasing and satisfies

$$\lim_{s \to -\infty} \gamma(s) = -1, \quad \lim_{s \to \infty} \gamma(s) = 1$$

exponentially and

$$\gamma(-s) = -\gamma(s) \quad \text{for every} \quad s \in \mathbb{R}.$$

Proof: Let $v : \mathbb{R} \to [-1, 1]$, $v' \in L^2(\mathbb{R})$, $\lim_{s \to -\infty} v(s) = -1$, $\lim_{s \to \infty} v(s) = 1$. Then,

$$K^1(v, \mathbb{R}) = \int_{-\infty}^{\infty} (v')^2 + W(v) dx \geq 2 \int_{-\infty}^{\infty} \sqrt{W(v)} v' \, dx = 2 \int_{-1}^{1} \sqrt{W(u)} du. \tag{3.6}$$

Taking the infimum on the left hand side over all such functions v, the definition of the optimal profile (3.5) delivers

$$\bar{\sigma} \geq 2 \int_{-1}^{1} \sqrt{W(u)} du = A_0, \tag{3.7}$$

and in (3.6), equality holds if and only if $v' = \sqrt{W(v)}$ a.e. in $[-1, 1]$. Thus, consider the system

$$\begin{cases} u' = \sqrt{W(u)} \\ u(0) = 0. \end{cases} \tag{3.8}$$

From Lemma 2.1, we know that $\sqrt{W}$ is locally Lipschitz continuous, so that (3.8) has a unique solution γ (see [26], Chapter I, Theorem 3.1). Since -1, 1 are the constant solutions of $u' = \sqrt{W(u)}$, γ only takes values in $(-1, 1)$ and is defined on the whole real line (see [26], Chapter I, Theorems 2.1 and 3.1).

Furthermore, $\gamma' = \sqrt{W(\gamma)} > 0$ implies that γ is strictly increasing, and since γ is bounded, we deduce $\gamma'(s) \to 0$ for $s \to \pm\infty$, so that (3.8) yields $\lim_{s \to \pm\infty} \sqrt{W(\gamma(s))} = 0$. Since -1, 1 are the zeros of W and γ is strictly increasing, we deduce

$$\lim_{s \to \pm\infty} \gamma(s) = \pm 1.$$

To show that $\gamma \in L^2(\mathbb{R})$, we observe that

$$\gamma'' = \frac{1}{2\sqrt{W(\gamma)}} W'(\gamma)\gamma' = \frac{1}{2}W'(\gamma),$$

and setting $p = \gamma'$ we obtain

$$p'' = \frac{1}{2}W''(\gamma)p \geq cp$$

on $[x_0, \infty)$ for some $x_0 > 0$, $c > 0$, since $W''(1) > 0$ and $\lim_{t \to \infty} \gamma(t) = 1$. For every $R > x_0$ we define $q_R : [x_0, R] \to \mathbb{R}$ by

$$q_R(t) = p(x_0) \exp(\sqrt{c}(x_0 - t)) + p(R) \exp(\sqrt{c}(t - R)).$$

Then $q_R'' = cq_R$ and thus

$$(p - q_R)'' \geq c(p - q_R) \quad \text{on} \quad [x_0, R].$$

Furthermore, $p(x_0) - q_R(x_0) \leq 0$, $p(R) - q_R(R) \leq 0$, since $p = \gamma' \geq 0$, so that the maximum principle (see [24], Corollary 3.2) implies

$$(p - q_R)(t) \leq \max\{p(x_0) - q_R(x_0), p(R) - q_R(R), 0\} = 0,$$

and for every $R > x_0$ we obtain the estimate

$$0 \leq \gamma'(t) = p(t) \leq q_R(t) = p(x_0) \exp(\sqrt{c}(x_0 - t)) + p(R) \exp(\sqrt{c}(t - R))$$

for every $t \in [x_0, R]$. For an arbitrary $t \geq x_0$, we choose $R_0 > x_0$ with $t \in [x_0, R]$ for every $R \geq R_0$, and letting $R \to \infty$ in above inequality we deduce

$$0 \leq \gamma'(t) \leq p(x_0) \exp \sqrt{c}(x_0 - t) \leq C \exp(-\alpha t)$$

for every $t \geq x_0$ with positive constants C, α. The proof of the exponential convergence $\lim_{t \to \infty} \gamma'(t) = -1$ is done analogously, so that we conclude $\gamma' \in L^2(\mathbb{R})$. From the definition of the optimal profile (3.5) we now deduce, recalling (3.7) and the fact that $\gamma' = \sqrt{W(\gamma)}$ turns (3.6) into an equation,

$$\bar{\sigma} \leq K^1(\gamma, \mathbb{R}) = 2 \int_{-1}^{1} \sqrt{W(\gamma)} du = A_0 \leq \bar{\sigma},$$

so that $\bar{\sigma} = A_0$. As for the exponential convergence $\lim_{t \to \pm\infty} = \pm 1$, we use the exponential convergence of γ' to derive

$$0 \leq 1 - \gamma(t) = \int_{t}^{\infty} \gamma'(t) dt \leq C \int_{t}^{\infty} \exp(-\alpha t) dt = \frac{C}{\alpha} \exp(-\alpha t)$$

for $t \geq x_0$, and the estimate for the negative half-axis is done analogously. Finally, the skew symmetry of γ follows from the fact that due to the symmetry of W, $t \mapsto -\gamma(-t)$ is also a solution of (3.8). Since the system is uniquely solvable, we deduce $\gamma(t) = -\gamma(-t)$. $\square$

3.3 Proof of the Modica-Mortola Theorem

The following proposition is a key argument in the proof of the Modica-Mortola theorem.

Proposition 3.4 *Let $(\varepsilon_n)_{n \in \mathbb{N}} \subset \mathbb{R}^+$ with $\varepsilon_n \to 0$ and $(v_n)_{n \in \mathbb{N}} \subset H^1(-1, 1)$ be a sequence of functions $v_n : (-1, 1) \to [-1, 1]$. Assume there is constant $\bar{C} > 0$ such that $K^{\varepsilon_n}(v_n; (-1, 1)) \leq \bar{C}$ for all $n \in \mathbb{N}$. Then there is a subsequence (not relabelled) $(v_n)_{n \in \mathbb{N}}$ converging in $L^2(-1, 1)$ to a function $v \in BV((-1, 1), \{-1, 1\})$, and*

$$\#(Sv \cap (-1, 1)) \leq \frac{\bar{C}}{\bar{\sigma}},$$

where $\bar{\sigma}$ is given by (3.5).

Remark: $\#(Sv \cap (-1, 1))$ is the set of inner discontinuity points, i.e. jumps, of v in $(-1, 1)$.

Proof: From $K^{\varepsilon_n}(v_n; (-1, 1)) \leq \bar{C}$, we deduce

$$\frac{1}{\varepsilon_n} \int_{-1}^{1} W(v_n) dx \leq \int_{-1}^{1} \varepsilon_n(v_n')^2 + \frac{1}{\varepsilon_n} W(v_n) dx = K^{\varepsilon_n}(v_n; (-1, 1)) \leq \bar{C}$$

which implies that $\int_{-1}^{1} W(v_n)dx \to 0$ for $n \to \infty$. Since $|v_n| \leq 1$, the fundamental theorem for Young measures (cf. [7]) yields that there is a subsequence $(v_n)_{n \in \mathbb{N}}$ (not relabelled) which generates a Young measure $\nu \in \mathcal{Y}((-1,1);[-1,1])$, i.e.

$$\lim_{n \to \infty} \int_{-1}^{1} g(v_n(s))h(s)ds = \int_{-1}^{1}\int_{-1}^{1} g(x)d\nu_s(x)h(s)ds \tag{3.9}$$

for every $g \in C[-1,1]$, $h \in L^1(-1,1)$. Setting $h(s) = 1$, $g = W$, we obtain

$$\int_{-1}^{1}\int_{-1}^{1} W(x)d\nu_s(x)ds = \lim_{n \to \infty} \int_{-1}^{1} W(v_n(s))ds = 0\,,$$

which due to the positivity of the integrands implies that $\int_{-1}^{1} W(x)d\nu_s(x) = 0$ for a.e. $s \in (-1,1)$. Thus, since -1 and $+1$ are the only zeros of W and $W > 0$ holds elsewhere, $\text{supp}(\nu_s)\subset \{-1,1\}$ for a.e. $s \in (-1,1)$, and we find a Borel measurable function (cf. Section 2.5) $\mu : (-1,1) \to [0,1]$ such that

$$\nu_s = \mu(s)\delta_{-1} + (1 - \mu(s))\delta_1 \quad \text{for a.e.} \quad s \in (-1,1).$$

Now, let $I \subset (-1,1)$ be an interval such that neither $\mu(s) = 0$ for a.e. $s \in I$ nor $\mu(s) = 1$ for a.e. $s \in I$. Then for every $\delta > 0$, there exists $n_0 = n_0(\delta) \in \mathbb{N}$ such that for every $n \geq n_0$ we will find points $a_n, b_n \in I$ with $v_n(a_n) \leq -1 + \delta$, $v_n(b_n) \geq 1 - \delta$.

Indeed, if we assume that this is not the case, we find a $\delta > 0$ and a subsequence $(v_n)_{n \in \mathbb{N}}$ (not relabelled) such that for every $n \in \mathbb{N}$, we have $v_n(s) \in (-1 + \delta, 1]$ for a.e. $s \in I$ or $v_n(s) \in [-1, 1 - \delta)$ for a.e $s \in I$. Choosing a further subsequence (not relabelled) we may assume that one of these two cases holds for any $n \in \mathbb{N}$. We consider the first case, the other one being treated analogously.

We choose $h = \chi_I$ and a $g \in C[-1,1]$ such that $g(-1) = 1$ and $\text{supp}(g)\subset [-1, -1 + \delta]$. Since $v_n(s) \in (-1 + \delta, 1]$ a.e. in I we have $g(v_n(s)) = 0$ a.e. in I and we obtain

$$\int_{-1}^{1}\int_{-1}^{1} g(x)d\nu_s(x)h(s)ds = \lim_{n \to \infty} \int_{-1}^{1} g(v_n(s))h(s)ds = \lim_{n \to \infty} \int_I g(v_n(s))ds = 0\,.$$

Since $\nu_s = \mu(s)\delta_{-1} + (1 - \mu(s))\delta_1$ and using $g(-1) = 1$, $g(1) = 0$, we deduce

$$0 = \int_{-1}^{1}\int_{-1}^{1} g(x)d\nu_s(x)h(s)ds = \int_I \int_{-1}^{1} g(x)d\nu_s(x)ds =$$

$$= g(-1) \int_I \mu(s)ds + g(1) \int_I (1 - \mu(s))ds = \int_I \mu(s)ds > 0\,,$$

which is a contradiction.

Now, let C_1 be the constant in Lemma 3.2, and choose $\delta > 0$ such that $C_1\delta\left(\frac{\bar{C}}{\bar{\sigma}} + 1\right) \leq \frac{\bar{\sigma}}{2}$. From what we have just shown, we can choose $n_0 \in \mathbb{N}$ such that for every $n \geq n_0$ we can find points $a_n, b_n \in I$ with $v_n(a_n) \leq -1 + \delta$, $v_n(b_n) \geq 1 - \delta$, and Lemma 3.2 yields

$$K^{\varepsilon_n}(v_n; I) \geq \bar{\sigma} - C_1\delta \quad \text{for all} \quad n \geq n_0\,.$$

Thus, there can exist at most $\frac{\bar{C}}{\bar{\sigma}}$ disjoint intervals of this type. If there were $\frac{\bar{C}}{\bar{\sigma}} + 1$ such intervals, we would deduce from the previous estimate and the choice of δ

$$\bar{C} \geq K^{\varepsilon_n}(v_n; (-1, 1)) \geq \left(1 + \frac{\bar{C}}{\bar{\sigma}}\right)(\bar{\sigma} - C_1 \delta) \geq \bar{\sigma} + \bar{C} - \frac{\bar{\sigma}}{2} = \bar{C} + \frac{\bar{\sigma}}{2}$$

which cannot be true due to $\bar{\sigma} = A_0 > 0$. In particular, we only have a finite number of disjoint intervals of the type mentioned above, which clearly implies that $\mu(s) \in \{0, 1\}$ for a.e. $s \in (-1, 1)$, and μ has at most $\frac{\bar{C}}{\bar{\sigma}}$ jumps, i.e. $\mu \in BV((-1, 1), \{0, 1\})$. We define a function $v \in BV((-1, 1), \{-1, 1\})$ by

$$v(s) = \begin{cases} 1, & \mu(s) = 0 \\ -1, & \mu(s) = 1 \end{cases} .$$

Then $v \in BV((-1, 1), \{-1, 1\})$, and

$$\#(Sv \cap (-1, 1)) \leq \frac{\bar{C}}{\bar{\sigma}}$$

from what we have shown for μ. We set $\Omega^+ = \{s \in (-1, 1) | v(s) = 1\} = \{s \in (-1, 1) | \mu(s) = 0\}$, $h = \chi_{\Omega^+}$ and $g(x) = x$. Then (3.9) and $\nu_s = \mu(s)\delta_{-1} + (1 - \mu(s))\delta_1$ yield

$$\lim_{n \to \infty} \int_{\Omega^+} v_n(s)dx = \int_{\Omega^+} \int_{-1}^{1} x d\nu_s(x)ds = -\int_{\Omega^+} \mu(s)ds + \int_{\Omega^+} (1 - \mu(s))ds = \lambda^1(\Omega^+),$$

where λ^1 denotes the Lebesgue measure on $\mathbb{R}$. Setting $\Omega^- = \{s \in (-1, 1) | v(s) = -1\}$, we analogously obtain

$$\lim_{n \to \infty} \int_{\Omega^-} v_n(s)ds = -\lambda^1(\Omega^-).$$

Thus, using $|v_n| \leq 1$ and the definitions of v, Ω^+ and Ω^-,

$$\int_{-1}^{1} |v_n(s) - v(s)|ds = \int_{\Omega^+} (1 - v_n(s))ds + \int_{\Omega^-} (1 + v_n(s))ds =$$

$$= \lambda^1(\Omega^+) - \int_{\Omega^+} v_n(s)ds + \lambda^1(\Omega^-) + \int_{\Omega^-} v_n(s)ds \to 0$$

for $n \to \infty$, hence $\|v_n - v\|_{L^1(-1,1)} \to 0$. Since $|v|, |v_n| \leq 1$, this yields

$$\int_{-1}^{1} |v_n(s) - v(s)|^2 ds \leq \|v_n - v\|_{L^\infty(-1,1)} \|v_n - v\|_{L^1(-1,1)} \leq 2\|v_n - v\|_{L^1(-1,1)} \to 0,$$

i.e. $v_n \to v$ in $L^2(-1, 1)$, and the proof is complete. $\qquad\qquad\square$

Now we are able to prove Theorem 3.1.

Proof of Theorem 3.1: To prove (i), we take sequences $(\varepsilon_n)_{n \in \mathbb{N}} \subset \mathbb{R}^+$ and $(v_n)_{n \in \mathbb{N}} \subset H^{1,0}_{per}(-1, 1)$ such that $\varepsilon_n \to 0$ and $(J^{\varepsilon_n}(v_n))_{n \in \mathbb{N}}$ is bounded, i.e. there exists a constant

$C > 0$ with $J^{\varepsilon n}(v_n) \leq C$ for every $n \in \mathbb{N}$. As in the proof of Lemma 3.2, we immediately deduce $\int_{-1}^{1} W(v_n) dx \to 0$ for $n \to \infty$.

Setting $w_n = (v_n \wedge 1) \vee (-1)$ for every $n \in \mathbb{N}$, we obtain a sequence of functions $(w_n)_{n \in \mathbb{N}} \subset H^1(-1, 1)$, $w_n : (-1, 1) \to [-1, 1]$. Define

$$\Omega_1 = \{ s \in (-1, 1) | \, |v_n(s)| \leq 1 \}, \quad \Omega_2 = \{ s \in (-1, 1) | \, |v_n(s)| > 1 \}.$$

Then $w_n(s) \in \{-1, 1\}$ for every $s \in \Omega_2$, which implies $W(w_n(s)) = 0$, $w_n'(s) = 0$ for a.e. $s \in \Omega_2$. Furthermore $w_n(s) = v_n(s)$, $w_n'(s) = v_n'(s)$ on Ω_1, which yields

$$K^{\varepsilon n}(w_n; (-1, 1)) = \int_{-1}^{1} \varepsilon(v_n')^2 + \frac{1}{\varepsilon} W(v_n) dx = \int_{\Omega_1} \varepsilon(v_n')^2 + \frac{1}{\varepsilon} W(v_n) dx \leq J^{\varepsilon n}(v_n) \leq C.$$

From Proposition 3.4 we deduce that there is a subsequence of $(w_n)_{n \in \mathbb{N}}$ (not relabelled) converging in $L^2(-1, 1)$ to a function $v \in BV((-1, 1), \{-1, 1\})$.

Using the growth properties of W we can prove that $\|v_n - v\|_{L^2(-1,1)} \to 0$ for $n \to \infty$. To do so, we first observe that

$$\|v_n - v\|_{L^2(-1,1)} \leq \|v_n - w_n\|_{L^2(-1,1)} + \|w_n - v\|_{L^2(-1,1)},$$

and since $w_n \to v$ in $L^2(-1, 1)$, we only need to show that $\|v_n - w_n\|_{L^2(-1,1)} \to 0$. Recalling $v_n = w_n$ on Ω_1, we obtain

$$\|v_n - w_n\|_{L^2(-1,1)}^2 = \int_{\Omega_2} (v_n - w_n)^2 ds = \int_{\Omega_2^+} (v_n - 1)^2 ds + \int_{\Omega_2^-} (v_n + 1)^2 ds,$$

where

$$\Omega_2^+ = \{ s \in \Omega_2 | v_n(s) > 1 \}, \quad \Omega_2^- = \{ s \in \Omega_2 | v_n(s) < -1 \}.$$

We will only consider the integral over Ω_2^+, the convergence of the other part is shown in an analogous way. For $\delta > 0$ arbitrarily small, we choose $\gamma = \gamma(\delta) > 0$ such that $\gamma^2 < \frac{\delta}{4}$. Due to the conditions on W, we find a $u_0 > 0$ such that $(u - 1)^2 \leq u^2 \leq CW(u)$ for every $u \geq u_0$, and since the function

$$u \mapsto \frac{W(u)}{(u-1)^2}$$

is continuous and positive on $[1 + \gamma, u_0]$, we find a $C_\gamma > 0$ such that $(u - 1)^2 \leq C_\gamma W(u)$ for $u \in [1 + \gamma, u_0]$. Thus, there exists a $C_\gamma > 0$ such that $(u - 1)^2 \leq C_\gamma W(u)$ for all $u \geq 1 + \gamma$ and we obtain

$$\int_{\Omega_2^+} (v_n - 1)^2 ds = \int_{\{v_n \in (1, 1+\gamma]\}} (v_n - 1)^2 ds + \int_{\{v_n > 1+\gamma\}} (v_n - 1)^2 ds$$

$$\leq 2\gamma^2 + C_\gamma \int_{\{v_n \geq 1+\gamma\}} W(v_n) ds \leq \frac{\delta}{2} + C_\gamma \int_{-1}^{1} W(v_n) ds.$$

From $\int_{-1}^{1} W(v_n) ds \to 0$, we deduce

$$\int_{\Omega_2^+} (v_n - 1)^2 ds < \delta$$

for $n \geq n_0(\delta)$. Using the analogous assertion for Ω_2^-, we get

$$\lim_{n \to \infty} \int_{\Omega_2} (v_n - w_n)^2 ds = 0,$$

which implies $v_n \to v$ in $L^2(-1, 1)$. Finally, as $(v_n)_{n \in \mathbb{N}} \subset X$, and thus $\int_{-1}^{1} v_n(s) ds = 0$ for every $n \in \mathbb{N}$, we immediately deduce $\int_{-1}^{1} v(s) ds = 0$, i.e. $v \in X$, from $v_n \to v$ in $L^2(-1, 1)$.

Thus we have shown that for every $(\varepsilon_n)_{n \in \mathbb{N}} \subset \mathbb{R}^+$ and $(v_n)_{n \in \mathbb{N}} \subset H^{1,0}_{per}(-1, 1)$ such that $\varepsilon_n \to 0$ and $(J^{\varepsilon_n}(v_n))_{n \in \mathbb{N}} \leq C$ for every $n \in \mathbb{N}$, there is a subsequence of (v_n) converging in X to a function $v \in X \cap BV((-1, 1), \{-1, 1\})$, and statement (i) follows easily.

As for statement (ii), let $(v_\varepsilon)_{\varepsilon > 0} \subset H^{1,0}_{per}(-1, 1)$ and $v \in X$ with $\lim_{\varepsilon \to 0} \|v_\varepsilon - v\|_{L^2(-1,1)} = 0$. We may assume $\liminf_{\varepsilon \to 0} J^\varepsilon(v_\varepsilon) < \infty$ since otherwise there is nothing to show, and we can pick a countable subsequence $(\varepsilon_n)_{n \in \mathbb{N}} \subset \mathbb{R}^+$ such that

$$\liminf_{n \to \infty} J^\varepsilon(v_\varepsilon) = \lim_{n \to \infty} J^{\varepsilon_n}(v_{\varepsilon_n}).$$

Thus, for $\delta > 0$ arbitrarily small, we find an $n_0 \in \mathbb{N}$ such that

$$J^{\varepsilon_n}(v_{\varepsilon_n}) \leq \liminf_{\varepsilon \to 0} J^\varepsilon(v_\varepsilon) + \delta =: C_\delta \quad \text{for } n \geq n_0 \, .$$

Setting $w_n = (v_{\varepsilon_n} \wedge 1) \vee (-1)$ as in the proof of statement (i), we obtain a sequence $(w_n)_{n \in \mathbb{N}} \subset H^1(-1, 1)$, $w_n : (-1, 1) \to [-1, 1]$ satisfying $K^{\varepsilon_n}(w_n; (-1, 1)) \leq J^{\varepsilon_n}(v_{\varepsilon_n}) \leq C_\delta$ for $n \geq n_0$. From Proposition 3.4, we deduce that (w_n) has a subsequence (not relabelled) converging in $L^2(-1, 1)$ to a function $w \in BV((-1, 1), \{-1, 1\})$, and the number of jumps of w in $(-1, 1)$ can be estimated from above by $\frac{C_\delta}{\sigma}$, i.e.

$$\#(Sw \cap (-1, 1)) \leq \frac{C_\delta}{\bar{\sigma}} = \frac{C_\delta}{A_0}$$

due to Lemma 3.3. Like in the proof of statement (i), we can show that $v_{\varepsilon_n} \to w$ for $n \to \infty$ in $L^2(-1, 1)$. Then $\int_{-1}^{1} v_{\varepsilon_n}(x) dx = 0$ yields $\int_{-1}^{1} w(x) dx = 0$, i.e. $w \in X \cap BV((-1, 1), \{-1, 1\})$. On the other hand, we know that $v_{\varepsilon_n} \to v$ in $L^2(-1, 1)$, which implies $v = w$ in $L^2(-1, 1)$, so that we can assume $v \in X \cap BV((-1, 1), \{-1, 1\})$ with

$$\#(Sv \cap (-1, 1)) \leq \frac{C_\delta}{A_0},$$

which yields

$$A_0 \#(Sv \cap (-1, 1)) \leq C_\delta = \liminf_{\varepsilon \to 0} J^\varepsilon(v_\varepsilon) + \delta \, ,$$

and since δ is arbitrary, we deduce

$$A_0 \#(Sv \cap (-1, 1)) \leq \liminf_{\varepsilon \to 0} J^\varepsilon(v_\varepsilon) \, .$$

Thus, if the periodic extension of v on $\mathbb{R}$ does not have a jump at -1, then the left-hand side is exactly $J(v)$, and statement (ii) follows in this case. If the periodic extension (now also denoted by v) has a jump at -1, we replace v by a suitable translation $T_\tau v = v(\cdot - \tau) \in BV((-1,1), \{-1,1\})$, $\tau \in \mathbb{R}$, such that

$$A_0 \#(ST_\tau v \cap (-1,1)) = A_0 \#(Sv \cap [-1,1)),$$

which means that the jump on the border is shifted into $(-1,1)$ (see Figure 3.3).

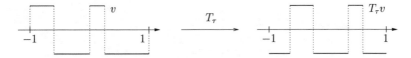

Figure 3.3: Shifting the jumps into $(-1,1)$

Replacing the sequence (v_ε) by $(T_\tau v_\varepsilon)$ (in the sense of periodic extension) we have $(T_\tau v_\varepsilon) \subset H^{1,0}_{per}(-1,1)$ and $\|T_\tau v_\varepsilon - T_\tau v\|_{L^2(-1,1)} \to 0$, and by applying what we have already shown we deduce

$$J(v) = A_0 \#(Sv \cap [-1,1)) = A_0 \#(ST_\tau v \cap (-1,1)) \le \liminf_{\varepsilon \to 0} J^\varepsilon(T_\tau v_\varepsilon) = \liminf_{\varepsilon \to 0} J^\varepsilon(v_\varepsilon),$$

where the last step simply follows from the translation invariance of J^ε. Thus, the proof of statement (ii) is complete.

To show (iii), we first assume that $v \in X \cap BV((-1,1), \{-1,1\})$ such that we find numbers

$$-1 = t_0 < t_1 < \cdots < t_N < t_{N+1} = 1,$$

with

$$v(x) = (-1)^i \quad \text{for} \quad x \in (t_{i-1}, t_i), \quad i = 1, \ldots, N+1,$$

where N is an even number, which means that v, extended periodically to $\mathbb{R}$, has no jump at the border of $(-1,1)$. In particular,

$$\#(Sv \cap [-1,1)) = \#(Sv \cap (-1,1)) = N.$$

Let $\gamma : \mathbb{R} \to (-1,1)$ be the optimal profile. In particular, $K^1(\gamma; \mathbb{R}) = \bar{\sigma} = A_0$ due to Lemma 3.3. We choose $\rho > 0$ such that the intervals $I_i = (t_i - \rho, t_i + \rho)$, $i = 1, \ldots, N$ satisfy $\bar{I}_i \cap \bar{I}_j = \emptyset$ for $i \ne j$ as well as $t_1 - \rho > -1$, $t_N + \rho < 1$. For every $\varepsilon > 0$ small enough, we define the function $v_\varepsilon \in H^1(-1,1)$ as follows:

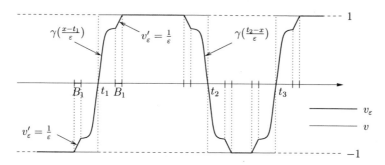

Figure 3.4: Definition of v_ε

More precisely,

$$v_\varepsilon(x) = (-1)^{i-1}\gamma(\frac{x-t_i}{\varepsilon}) \quad \text{for} \quad x \in I_i = (t_i - \rho, t_i + \rho), \quad i = 1, \dots, N$$

$$v_\varepsilon'(x) = (-1)^{i-1}\frac{1}{\varepsilon} \quad \text{for} \quad x \in B_i, \quad i = 1, \dots, N$$

$$v_\varepsilon(x) = v(x) \quad \text{elsewhere on } (-1,1).$$

Since γ is skew symmetric, we easily see that $\int_{I_i \cup B_i} v_\varepsilon = 0$, and since we also have $\int_{I_i \cup B_i} v = 0$ and v and v_ε coincide outside the unity of all I_i and B_i, we deduce

$$\int_{-1}^{1} v_\varepsilon(x)dx = \int_{-1}^{1} v(x)dx = 0,$$

so that $v_\varepsilon \in H_{per}^{1,0}(-1,1)$ (the periodicity easily following from the definitions of v, v_ε and the fact that N is even). Recalling that $\lim_{t\to\pm\infty} \gamma(t) = \pm1$ exponentially, we see that

$$v_\varepsilon(t_i \pm \rho) = \gamma(\pm\frac{\rho}{\varepsilon}) = \pm1 + O(\varepsilon), \quad i \text{ odd}$$

$$v_\varepsilon(t_i \pm \rho) = -\gamma(\pm\frac{\rho}{\varepsilon}) = \mp1 + O(\varepsilon), \quad i \text{ even},$$

and since the slope on B_i is ε^{-1}, the size of the sets B_i is of order $\leq O(\varepsilon^2)$. It is easy to see that $v_\varepsilon \to v$ pointwise for $\varepsilon \to 0$, which implies

$$\lim_{\varepsilon\to 0} \|v_\varepsilon - v\|_{L^2(-1,1)} = 0.$$

Furthermore, the scaling property (3.4) implies

$$K^\varepsilon(v_\varepsilon; I_i) = K^1(\gamma; (-\frac{\rho}{\varepsilon}, \frac{\rho}{\varepsilon})) \leq K^1(\gamma; \mathbb{R}) = A_0$$

due to Lemma 3.3. On the other hand, using $|B_i| \leq O(\varepsilon^2)$, $v_\varepsilon' = \pm\frac{1}{\varepsilon}$ and $W(v) \leq C$,

$$K^\varepsilon(v_\varepsilon; B_i) = \int_{B_i} \varepsilon(v_\varepsilon')^2 + \frac{1}{\varepsilon}W(v)dx \leq \frac{1}{\varepsilon}|B_i| + \frac{C}{\varepsilon}|B_i| \leq C\varepsilon,$$

and since $v_\varepsilon = v$ outside the union of all I_i and B_i, so that $v'_\varepsilon = 0$, $W(v_\varepsilon) = W(\pm 1) = 0$ holds there, we obtain

$$J^\varepsilon(v_\varepsilon) = \sum_{i=1}^{N} K^\varepsilon(v_\varepsilon; I_i) + \sum_{i=1}^{N} K^\varepsilon(v_\varepsilon; B_i) \leq A_0 N + CN\varepsilon = (A_0 + C\varepsilon)\#(Sv \cap [-1,1)),$$

which implies
$$\limsup_{\varepsilon \to 0} J^\varepsilon(v_\varepsilon) \leq A_0 \#(Sv \cap [-1,1)),$$

so that we have shown (iii) for the special case. In the case where the periodic extension of v has a jump on the border of $(-1,1)$, we have to proceed as in the proof of (ii) and replace (the periodic extension of) v by a suitable translation $T_\tau v$ such that

$$\#(ST_\tau v \cap (-1,1)) = \#(Sv \cap [-1,1)),$$

i.e. $T_\tau v$ does not have a corner on the border of $(-1,1)$ (or, the corner -1 of v is shifted into $(-1,1)$). Applying what we have shown to $T_\tau v$, we obtain a sequence $(w_\varepsilon)_{\varepsilon>0} \subset H^{1,0}_{per}(-1,1)$ with $\|w_\varepsilon - T_\tau v\|_{L^2(-1,1)} \to 0$ for $\varepsilon \to 0$ and

$$\limsup_{\varepsilon \to 0} J^\varepsilon(w_\varepsilon) \leq A_0 \#(ST_\tau v \cap [-1,1)) = A_0 \#(ST_\tau v \cap (-1,1)),$$

and setting $v_\varepsilon = T_{-\tau} w_\varepsilon$ for every $\varepsilon > 0$ we obtain a sequence $(v_\varepsilon)_{\varepsilon>0} \subset H^{1,0}_{per}(-1,1)$ with $\|v_\varepsilon - v\|_{L^2(-1,1)} = \|w_\varepsilon - T_\tau v\|_{L^2(-1,1)} \to 0$ with

$$\limsup_{\varepsilon \to 0} J^\varepsilon(v_\varepsilon) = \limsup_{\varepsilon \to 0} J^\varepsilon(w_\varepsilon) \leq A_0 \#(ST_\tau v \cap (-1,1)) = A_0 \#(Sv \cap [-1,1)),$$

which completes the proof of Theorem 3.1. $\qquad\square$

Chapter 4

The Γ-Limit of I^ε

4.1 Application of the Modica-Mortola Theorem

By applying the version of the Modica-Mortola theorem shown in the previous chapter, we will identify the Γ-limit I of the functionals I^ε (extended by $+\infty$ to a larger space) with respect to the $H^1(-1,1)$ topology. Once this is done, we show how local minimizers of I^ε for $\varepsilon > 0$ small can be characterized with the help of local minimizers of the Γ-limit I. Corollary 4.9 reduces this simpler problem to a local minimum property for the non-local energy E on the set of periodic "sawtooth functions" u with zero average, $u' \in \{\pm 1\}$ and a fixed number of corners. This turns out to be a finite-dimensional problem.

For every $\varepsilon > 0$, we extend I^ε by $+\infty$ to a larger set Y as follows:

$$I^\varepsilon(u) = \begin{cases} \displaystyle\int_{-1}^{1} \varepsilon(u'')^2 + \frac{1}{\varepsilon}W(u')dx + E(u) & \text{for } u \in H^{2,0}_{per}(-1,1) \\ \infty & \text{for } u \in Y \setminus H^{2,0}_{per}(-1,1), \end{cases} \tag{4.1}$$

where Y is the Hilbert space

$$Y = H^{1,0}_{per}(-1,1) = \{u \in H^1(-1,1) \,|\, u(-1) = u(1), \int_{-1}^{1} u(x)dx = 0\}$$

endowed with norm $\|\cdot\|_Y = \|\cdot\|_{H^1(-1,1)}$ and E is the nonlocal energy given by

$$E(u) = \int_{-1}^{1}\int_{-1}^{1} g(x-y)(u(x)-u(y))^2 dy dx. \tag{4.2}$$

We define the set $\mathcal{S}(-1,1)$ of "sawtooth functions" whose distributional derivative u' takes values -1 and 1 only and has finitely many jumps, which are the corners of u:

$$\mathcal{S}(-1,1) = \{u : [-1,1] \to \mathbb{R} \,|\, u' \in BV((-1,1),\{-1,1\})\}. \tag{4.3}$$

Obviously, $Y \cap \mathcal{S}(-1,1)$ is given by

$$\mathcal{S}^0_{per}(-1,1) = \{u \in \mathcal{S}(-1,1) \,|\, u(-1) = u(1), \int_{-1}^{1} u(x)dx = 0\}, \tag{4.4}$$

which is the set of periodic sawtooth functions with slope $u' = \pm 1$, a finite number of corners and zero average. We define a functional $I : Y \to [0, \infty]$ by

$$I(u) = \begin{cases} A_0 \#(Su' \cap [-1, 1)) + E(u) & \text{for } u \in \mathcal{S}^0_{per}(-1, 1) \\ \\ \infty & \text{for } u \in Y \setminus \mathcal{S}^0_{per}(-1, 1), \end{cases} \tag{4.5}$$

where

$$A_0 = 2 \int_{-1}^1 \sqrt{W(v)} dv$$

as before. Here, $\#(Su' \cap [-1, 1))$ counts the numbers of corners on $[-1, 1)$ in that sense that u by definition has a corner at -1 if its periodic extension to $\mathbb{R}$ has. With these definitions, we can show the following:

Theorem 4.1   *(i) Let $(\varepsilon_n)_{n \in \mathbb{N}} \subset \mathbb{R}^+$ such that $\varepsilon_n \to 0$ for $n \to \infty$, and $(u_n)_{n \in \mathbb{N}} \subset H^{2,0}_{per}(-1, 1)$ such that $(I^{\varepsilon_n}(u_n))_{n \in \mathbb{N}}$ is a bounded sequence in $\mathbb{R}$. Then $(u_n)_{n \in \mathbb{N}}$ is relatively compact in Y, and every Y-cluster point of $(u_n)_{n \in \mathbb{N}}$ lies in $\mathcal{S}^0_{per}(-1, 1)$.*

(ii) Let $(u_\varepsilon)_{\varepsilon > 0} \subset H^{2,0}_{per}(-1, 1)$, $u \in Y$ such that $\|u_\varepsilon - u\|_{H^1(-1,1)} \to 0$ for $\varepsilon \to 0$. Then

$$I(u) \le \liminf_{\varepsilon \to 0} I^\varepsilon(u_\varepsilon).$$

(iii) For every $u \in \mathcal{S}^0_{per}(-1, 1)$, we find a sequence $(u_\varepsilon)_{\varepsilon > 0} \subset H^{2,0}_{per}(-1, 1)$ such that $\|u_\varepsilon - u\|_{H^1(-1,1)} \to 0$ for $\varepsilon \to 0$ and

$$\limsup_{\varepsilon \to 0} I^\varepsilon(u_\varepsilon) \le I(u).$$

In particular,

$$I^\varepsilon \overset{\Gamma}{\to} I \quad in \quad Y.$$

Apart from the Modica-Mortola-theorem, we need the following Lemma to show Theorem 4.1:

Lemma 4.2 *The nonlocal energy E defined by (4.2), viewed as a functional*

$$E : L^2(-1, 1) \to \mathbb{R},$$

is continuous. In particular, E is continuous with respect to $\| \cdot \|_{H^1(-1,1)}$.

Proof of Lemma 4.2: Let $(u_n)_{n \in \mathbb{N}} \subset L^2(-1, 1)$, $u \in L^2(-1, 1)$ such that $u_n \to u$ in $L^2(-1, 1)$ for $n \to \infty$. Using $|g| \le g_1$ and the Cauchy-Schwarz inequality, we obtain

$$|E(u_n) - E(u)| = \left| \int_{-1}^1 \int_{-1}^1 g(x - y)[(u_n(x) - u_n(y))^2 - (u(x) - u(y))^2] dy dx \right|$$

$$\leq \; g_1 \int_{-1}^{1} \int_{-1}^{1} \left| (u_n(x) - u_n(y))^2 - (u(x) - u(y))^2 \right| dy\,dx$$

$$\leq \; g_1 \left[\int_{-1}^{1} \int_{-1}^{1} \left[(u_n(x) - u_n(y)) - (u(x) - u(y)) \right]^2 dy\,dx \right]^{\frac{1}{2}} \times$$

$$\times \left[\int_{-1}^{1} \int_{-1}^{1} \left[(u_n(x) - u_n(y)) + (u(x) - u(y)) \right]^2 dy\,dx \right]^{\frac{1}{2}}$$

$$= \; g_1 \left[\int_{-1}^{1} \int_{-1}^{1} \left[(u_n(x) - u(x)) - (u_n(y) - u(y)) \right]^2 dy\,dx \right]^{\frac{1}{2}} \times$$

$$\times \left[\int_{-1}^{1} \int_{-1}^{1} \left[(u_n(x) + u(x)) - (u_n(y) + u(y)) \right]^2 dy\,dx \right]^{\frac{1}{2}}$$

$$= \; g_1 \left[4 \int_{-1}^{1} (u_n(x) - u(x))^2 dx - 2 \left[\int_{-1}^{1} (u_n(x) - u(x)) dx \right]^2 \right]^{\frac{1}{2}} \times$$

$$\times \left[4 \int_{-1}^{1} (u_n(x) + u(x))^2 dx - 2 \left[\int_{-1}^{1} (u_n(x) + u(x)) dx \right]^2 \right]^{\frac{1}{2}}$$

$$\leq \; g_1 \left[4 \int_{-1}^{1} (u_n(x) - u(x))^2 dx \right]^{\frac{1}{2}} \left[4 \int_{-1}^{1} (u_n(x) + u(x))^2 dx \right]^{\frac{1}{2}}$$

$$\leq \; C \| u_n - u \|_{L^2(-1,1)} \| u_n + u \|_{L^2(-1,1)} \leq C \| u_n - u \|_{L^2(-1,1)} \,,$$

since the convergence of $(u_n)_{n \in \mathbb{N}}$ implies the boundedness of $\| u_n \|_{L^2(-1,1)}$ in $\mathbb{R}$, and the proof is complete. $\qquad \square$

Proof of Theorem 4.1: We first remark that

$$I^\varepsilon(u) = I_1^\varepsilon(u) + E(u), \quad I(u) = I_1(u) + E(u) \ \text{ for } \ u \in Y \,,$$

where the functionals $I_1^\varepsilon : Y \to [0, \infty]$, $\varepsilon > 0$ and $I_1 : Y \to [0, \infty]$ are defined by

$$I_1^\varepsilon(u) = \begin{cases} \displaystyle \int_{-1}^{1} \varepsilon(u'')^2 + \frac{1}{\varepsilon} W(u') dx & \text{for } u \in H_{per}^{2,0}(-1,1) \\ \infty & \text{for } u \in Y \setminus H_{per}^{2,0}(-1,1) . \end{cases}$$

$$I_1(u) = \begin{cases} A_0 \#(Su' \cap [-1,1)) & \text{for } u \in \mathcal{S}_{per}^0(-1,1) \\ \infty & \text{for } u \in Y \setminus \mathcal{S}_{per}^0(-1,1) , \end{cases}$$

Also, we need to recall the definitions of the functionals $J^\varepsilon, J : X \to [0, \infty]$ given by (3.2) and (3.3), respectively.

To show (ii), let $(u_\varepsilon)_{\varepsilon > 0} \subset H_{per}^{2,0}(-1,1)$, $u \in Y$ such that $\| u_\varepsilon - u \|_{H^1(-1,1)} \to 0$ for $\varepsilon \to 0$. We may assume $\liminf_{\varepsilon \to 0} I^\varepsilon(u_\varepsilon) < \infty$, since otherwise there is nothing to show. In particular,

$$\liminf_{\varepsilon \to 0} I_1^\varepsilon(u_\varepsilon) < \infty.$$

For every $\varepsilon > 0$, we have $u'_\varepsilon \in H^1(-1,1)$ with $u'_\varepsilon(-1) = u'_\varepsilon(1)$, and since

$$\int_{-1}^{1} u'_\varepsilon(x)dx = u_\varepsilon(1) - u_\varepsilon(-1) = 0,$$

we obtain $(u'_\varepsilon)_{\varepsilon>0} \subset H^{1,0}_{per}(-1,1)$ with $J^\varepsilon(u'_\varepsilon) = I^\varepsilon_1(u_\varepsilon)$ for every $\varepsilon > 0$. Obviously, we have $\|u'_\varepsilon - u'\|_{L^2(-1,1)} \to 0$, which together with $\int_{-1}^{1} u'_\varepsilon dx = 0$ implies $\int_{-1}^{1} u'dx = 0$, i.e. $u' \in X$. Using Theorem 3.1(ii), we obtain

$$J(u') \leq \liminf_{\varepsilon \to 0} J^\varepsilon(u'_\varepsilon) = \liminf_{\varepsilon \to 0} I^\varepsilon_1(u_\varepsilon).$$

Since the right-hand side is finite, we conclude $u' \in X \cap BV((-1,1),\{-1,1\})$, which implies $u \in \mathcal{S}(-1,1)$. Thus $u \in Y \cap \mathcal{S}(-1,1) = \mathcal{S}^0_{per}(-1,1)$ with $I_1(u) = J(u')$, and

$$I_1(u) \leq \liminf_{\varepsilon \to 0} I^\varepsilon_1(u_\varepsilon).$$

Since $\|u_\varepsilon - u\|_{H^1(-1,1)} \to 0$, Lemma 4.2 yields $\lim_{\varepsilon \to 0} E(u_\varepsilon) = E(u)$, and we obtain

$$I(u) = I_1(u) + E(u) \leq \liminf_{\varepsilon \to 0}(I^\varepsilon_1(u_\varepsilon) + E(u_\varepsilon)) = \liminf_{\varepsilon \to 0} I^\varepsilon(u_\varepsilon).$$

To prove (iii), let $u \in \mathcal{S}^0_{per}(-1,1)$. Then $u' \in BV((-1,1),\{-1,1\})$, and

$$\int_{-1}^{1} u'(x)dx = u(1) - u(-1) = 0.$$

Thus, $u' \in X \cap BV((-1,1),\{-1,1\})$, and $J(u') = I_1(u)$. Using Theorem 3.1(iii), we can choose a sequence $(v_\varepsilon)_{\varepsilon>0} \subset H^{1,0}_{per}(-1,1)$ with $\|v_\varepsilon - u'\|_{L^2(-1,1)} \to 0$ for $\varepsilon \to 0$ and

$$\limsup_{\varepsilon \to 0} J^\varepsilon(v_\varepsilon) \leq J(u') = I_1(u). \tag{4.6}$$

For every $\varepsilon > 0$, we define $u_\varepsilon : (-1,1) \to \mathbb{R}$ by

$$u_\varepsilon(x) = \int_{-1}^{x} v_\varepsilon(\xi)d\xi - \frac{1}{2}\int_{-1}^{1}\int_{-1}^{\eta} v_\varepsilon(\xi)d\xi d\eta.$$

Then $u_\varepsilon \in H^2(-1,1)$, and since $u'_\varepsilon = v_\varepsilon \in X$, we have $u_\varepsilon(1) - u_\varepsilon(-1) = \int_{-1}^{1} v_\varepsilon(x)dx = 0$, so that $u_\varepsilon(-1) = u_\varepsilon(1)$. Furthermore, $u'_\varepsilon(-1) = v_\varepsilon(-1) = v_\varepsilon(1) = u'_\varepsilon(1)$, and since

$$\int_{-1}^{1} u_\varepsilon(x)dx = 0$$

by definition of u_ε, we conclude $(u_\varepsilon)_{\varepsilon>0} \subset H^{2,0}_{per}(-1,1)$, and for every $\varepsilon > 0$ we have

$$J^\varepsilon(v_\varepsilon) = J^\varepsilon(u'_\varepsilon) = I^\varepsilon_1(u_\varepsilon). \tag{4.7}$$

To show that $u_\varepsilon \to u$ in $L^2(-1,1)$ for $\varepsilon \to 0$, we rewrite u as

$$u(x) = \int_{-1}^{x} u'(\xi)d\xi + c_u,$$

where $c_u \in \mathbb{R}$ is chosen so that the condition $\int_{-1}^{1} u(x)dx = 0$ holds, and a simple calculation yields

$$u(x) = \int_{-1}^{x} u'(\xi)d\xi - \frac{1}{2}\int_{-1}^{1}\int_{-1}^{\eta} u'(\xi)d\xi d\eta.$$

Thus for every $x \in (-1, 1)$

$$|u_\varepsilon(x) - u(x)| \leq \int_{-1}^{x} |v_\varepsilon(\xi) - u'(\xi)|d\xi + \frac{1}{2}\int_{-1}^{1}\int_{-1}^{\eta} |v_\varepsilon(\xi) - u'(\xi)|d\xi d\eta$$

$$\leq 2\int_{-1}^{1} |v_\varepsilon(\xi) - u'(\xi)|d\xi \leq 2\sqrt{2}\|v_\varepsilon - u'\|_{L^2(-1,1)}$$

due to the Cauchy-Schwarz inequality, and we deduce

$$\|u_\varepsilon - u\|_{L^2(-1,1)}^2 \leq 8\int_{-1}^{1} \|v_\varepsilon - u'\|_{L^2(-1,1)}^2 dx \leq 16\|v_\varepsilon - u'\|_{L^2(-1,1)}^2 \to 0$$

for $\varepsilon \to 0$. Thus, $\|u_\varepsilon - u\|_{L^2(-1,1)} \to 0$ for $\varepsilon \to 0$ and since we know that $\|v_\varepsilon - u'\|_{L^2(-1,1)} \to 0$ for $\varepsilon \to 0$ and that $v_\varepsilon = u'_\varepsilon$ for every $\varepsilon > 0$, this implies $\|u_\varepsilon - u\|_{H^1(-1,1)} \to 0$. Combining (4.6) and (4.7) we get

$$\limsup_{\varepsilon \to 0} I_1^\varepsilon(u_\varepsilon) = \limsup_{\varepsilon \to 0} J^\varepsilon(v_\varepsilon) \leq I_1(u),$$

and using Lemma 4.2 yields $E(u_\varepsilon) \to E(u)$ for $\varepsilon \to 0$, so that we obtain

$$\limsup_{\varepsilon \to 0} I^\varepsilon(u_\varepsilon) = \limsup_{\varepsilon \to 0}(I_1^\varepsilon(u_\varepsilon) + E(u_\varepsilon)) \leq I_1(u) + E(u) = I(u).$$

To prove (i), we take sequences $(\varepsilon_n)_{n\in\mathbb{N}} \subset \mathbb{R}^+$ such that $\varepsilon_n \to 0$ for $n \to \infty$ and $(u_n)_{n\in\mathbb{N}} \subset H_{per}^{2,0}(-1,1)$ such that $(I^{\varepsilon_n}(u_n))_{n\in\mathbb{N}}$ is a bounded sequence in $\mathbb{R}$. Then $(I_1^{\varepsilon_n}(u_n))_{n\in\mathbb{N}}$ is also bounded. For every $n \in \mathbb{N}$, we have $u'_n \in H^1(-1,1)$ with $u'_n(-1) = u'_n(1)$ and

$$\int_{-1}^{1} u'_n(x)dx = u_n(1) - u_n(-1) = 0$$

due to periodicity of u_n. Thus $(u'_n)_{n\in\mathbb{N}} \subset H_{per}^{1,0}(-1,1)$, and $J^{\varepsilon_n}(u'_n) = I_1^{\varepsilon_n}(u_n)$ for every $n \in \mathbb{N}$, so that $(J^{\varepsilon_n}(u'_n))_{n\in\mathbb{N}}$ is also a bounded sequence in $\mathbb{R}$. By Theorem 3.1 (i), we find a subsequence of $(u'_n)_{n\in\mathbb{N}}$ (not relabelled) and a $v \in X \cap BV((-1,1), \{-1,1\})$ such that $\|u'_n - v\|_{L^2(-1,1)} \to 0$ for $n \to \infty$. We define $u : (-1,1) \to \mathbb{R}$ by

$$u(x) = \int_{-1}^{x} v(\xi)d\xi - \frac{1}{2}\int_{-1}^{1}\int_{-1}^{\eta} v(\xi)d\xi d\eta.$$

Clearly, $u \in \mathcal{S}(-1,1)$ with $u' = v$, and since $v \in X$, we have $u(1) - u(-1) = \int_{-1}^{1} v(x)dx = 0$, i.e. $u(-1) = u(1)$. Also, we easily see that $\int_{-1}^{1} u(x)dx = 0$, so that $u \in \mathcal{S}_{per}^0(-1,1)$. Arguing as in the proof of (ii) and using the zero average condition for u_n, we can show that

$$u_n(x) = \int_{-1}^{x} u'_n(\xi)d\xi - \frac{1}{2}\int_{-1}^{1}\int_{-1}^{\eta} u'_n(\xi)d\xi d\eta$$

for every $n \in \mathbb{N}$, and $\|u_n - u\|_{L^2(-1,1)} \to 0$ follows analogously as in the proof of (ii) from $\|u'_n - v\|_{L^2(-1,1)} \to 0$ for $n \to \infty$. Since $u' = v$, we deduce $\|u_n - u\|_{H^1(-1,1)} \to 0$, and the proof is complete. $\square$

4.2   Γ-Limits and Local Minimizers

We will now use Theorem 4.1 to draw a connection between local minimizers of the functionals I^ε for small $\varepsilon > 0$ and those of I. This issue has been dealt with in literature on Γ-convergence (see Theorem 4.1 in [31] for a quite general statement, §3 in [17] and Proposition 2.3 in [41]). Since these theorems assume that the local minimizers of the Γ-limit are strict (isolated) ones, however, we can not apply them directly to our problem, since the translation invariance of I (in sense of periodic extension of functions to $\mathbb{R}$) implies that strict local minimizers cannot exist.

Indeed, the translation invariance implies that any $H^1(-1, 1)$-ball around a function $u \in \mathcal{S}^0_{per}(-1, 1)$ contains arbitrary small translations of u leading to the same energy $I(u)$. (The invariance under translations is obvious for the nonlocal term E. As for the number of corners $\#(Su' \cap [-1, 1))$, since a possible corner at -1 is counted, no "new" corners can appear under small shifts, which means that this term is also invariant under translations). A first step is to consider orbits of periodic functions rather than the functions themselves:

Definition 4.3 *Let $u \in Y$, and the periodic extension of u to $\mathbb{R}$ also be denoted by u. For an arbitrary $\tau \in \mathbb{R}$, we define the τ-**translation** $T_\tau u : \mathbb{R} \to \mathbb{R}$ by*

$$T_\tau u(x) = u(x - \tau)$$

*for every $x \in \mathbb{R}$, and the **orbit of u** is defined as the set*

$$\mathcal{O}(u) = \{T_\tau u \mid \tau \in \mathbb{R}\}$$

of all translations of u. For $A \subset Y$, we define

$$\mathcal{O}(A) = \bigcup_{u \in A} \mathcal{O}(u) = \{u \in Y \mid \exists \tau \in \mathbb{R} : T_\tau u \in A\}.$$

Remarks:

 a) It is obvious that $u \in Y$ implies $T_\tau u \in Y$ (in sense of restriction to $(-1, 1)$).

 b) The set equation follows from $v = T_\tau u \Leftrightarrow u = T_{-\tau} v$.

Definition 4.4 *For $u \in Y$ and $\delta > 0$, we define*

$$B_\delta(u) = \{v \in Y \mid \|v - u\|_{H^1(-1,1)} < \delta\}$$

as the open H^1-ball around u with radius δ, and for a set $A \subset Y$, we define

$$B_\delta(A) = \{u \in Y \mid \exists v \in A : u \in B_\delta(v)\}.$$

We will make use of the following simple property:

Lemma 4.5 *For every $u \in Y$, $\delta > 0$, we have*

$$B_\delta(\mathcal{O}(u)) = \mathcal{O}(B_\delta(u)).$$

Proof: Let $v \in B_\delta(\mathcal{O}(u))$, then we find $w \in \mathcal{O}(u)$ such that $\|v - w\|_{H^1(-1,1)} < \delta$. Hence there exists $\tau \in \mathbb{R}$ with $w = T_\tau u$, and we have $\|T_{-\tau}v - u\|_{H^1(-1,1)} = \|v - T_\tau u\|_{H^1(-1,1)} = \|v - w\|_{H^1(-1,1)} < \delta$. Thus $T_{-\tau}v \in B_\delta(u)$, and we deduce $v \in \mathcal{O}(B_\delta(u))$.
If $v \in \mathcal{O}(B_\delta(u))$, then $T_\tau v \in B_\delta(u)$ for some $\tau \in \mathbb{R}$, and we get $\|v - T_{-\tau}u\|_{H^1(-1,1)} = \|T_\tau v - u\|_{H^1(-1,1)} < \delta$, thus $v \in B_\delta(\mathcal{O}(u))$. $\qquad\square$

The following theorem shows how we can obtain local minimizers of the functionals I^ε from local minimizers of the Γ-limit I. It is based on Proposition 2.3 in [41] which we transfer to the translation invariant case, and the proof is adopted from the one given therein.

Theorem 4.6 *Let $\bar{u} \in \mathcal{S}_{per}^0(-1,1)$. Assume that there exists $\delta > 0$ such that for every $u \in \mathcal{S}_{per}^0(-1,1)$ with $\|u - \bar{u}\|_{H^1(-1,1)} < \delta$ and $u \notin \mathcal{O}(\bar{u})$ we have*

$$I(u) > I(\bar{u}).$$

Then there exists $\varepsilon_0 > 0$ such that for every $\varepsilon \in (0, \varepsilon_0)$ there is a $u_\varepsilon \in B_{\frac{\delta}{2}}(\mathcal{O}(\bar{u})) \cap H_{per}^{2,0}(-1,1)$ with

$$I^\varepsilon(u_\varepsilon) \le I^\varepsilon(u) \quad \text{for every} \quad u \in B_{\frac{\delta}{2}}(\mathcal{O}(\bar{u})) \cap H_{per}^{2,0}(-1,1).$$

Furthermore,

$$\inf_{u \in \mathcal{O}(\bar{u})} \|u_\varepsilon - u\|_{H^1(-1,1)} \to 0 \quad \text{for} \quad \varepsilon \to 0. \tag{4.8}$$

Remarks:

a) The assumption in the theorem means that $\bar{u}$ is a "local minimizer up to translation" in the sense that the only functions $u \in \mathcal{S}_{per}^0(-1,1)$ that are close to $\bar{u}$ in $H^1(-1,1)$ and satisfy $I(u) \le I(\bar{u})$ are translations of $\bar{u}$, hence $I(u) = I(\bar{u})$.

b) In the following, for a set of functions $A \subset Y$, $\bar{A}$ denotes the closure of A in the topology of Y, i.e. with respect to $\|\cdot\|_{H^1(-1,1)}$.

Proof: Let $\delta > 0$ be as described. For every $\varepsilon > 0$, we find a sequence $(u_{\varepsilon,n})_{n \in \mathbb{N}} \subset B_{\frac{\delta}{2}}(\mathcal{O}(\bar{u})) \cap H_{per}^{2,0}(-1,1)$ such that

$$\inf_{u \in B_{\frac{\delta}{2}}(\mathcal{O}(\bar{u}))} I^\varepsilon(u) = \lim_{n \to \infty} I^\varepsilon(u_{\varepsilon,n}) \tag{4.9}$$

(recall that I^ε is only finite on $H_{per}^{2,0}(-1,1)$). As in Section 2.2, we infer that for every $\varepsilon > 0$ we find $u_\varepsilon \in H_{per}^{2,0}(-1,1)$ and a subsequence of $(u_{\varepsilon,n})_{n \in \mathbb{N}}$ (not relabelled) such that

$$u_{\varepsilon,n} \xrightarrow[n \to \infty]{} u_\varepsilon \quad \text{in} \quad H_{per}^{2,0}(-1,1) \tag{4.10}$$

and

$$I^\varepsilon(u_\varepsilon) \le \liminf_{n\to\infty} I^\varepsilon(u_{\varepsilon,n})$$

holds. Using (4.9), we conclude

$$I^\varepsilon(u_\varepsilon) \le \liminf_{n\to\infty} I^\varepsilon(u_{\varepsilon,n}) = \lim_{n\to\infty} I^\varepsilon(u_{\varepsilon,n}) = \inf_{u\in B_{\frac{\delta}{2}}(\mathcal{O}(\bar u))} I^\varepsilon(u) \tag{4.11}$$

for every $\varepsilon > 0$. Since (4.10) holds with $(u_{\varepsilon,n})_{n\in\mathbb{N}} \subset B_{\frac{\delta}{2}}(\mathcal{O}(\bar u)) = \mathcal{O}(B_{\frac{\delta}{2}}(\bar u))$ (see Lemma 4.5), the compact imbedding $H_{per}^{2,0}(-1,1) \hookrightarrow H^1(-1,1)$ (see [1]) implies $u_\varepsilon \in \overline{\mathcal{O}(B_{\frac{\delta}{2}}(\bar u))}$ for every $\varepsilon > 0$.

We aim to show that $u_\varepsilon \in \mathcal{O}(B_\delta(\bar u))$ for ε sufficiently small. Assume that this is not the case. Then we find a sequence $(\varepsilon_l)_{l\in\mathbb{N}} \subset \mathbb{R}^+$ such that $\varepsilon_l \to 0$ for $l \to \infty$ and

$$u_{\varepsilon_l} \in \overline{\mathcal{O}(B_{\frac{\delta}{2}}(\bar u))} \setminus \mathcal{O}(B_{\frac{\delta}{2}}(\bar u)) \tag{4.12}$$

for every $l \in \mathbb{N}$, and we can choose $\tau_l \in [0,2]$, respectively, with

$$\|u_{\varepsilon_l} - T_{\tau_l}\bar u\|_{H^1(-1,1)} = \frac{\delta}{2} \tag{4.13}$$

for every $l \in \mathbb{N}$. Using Theorem 4.1 (iii), for every l we can choose $v_{\varepsilon_l} \in H_{per}^{2,0}(-1,1)$ such that $(v_{\varepsilon_l})_{l\in\mathbb{N}} \subset B_{\frac{\delta}{2}}(\bar u) \subset \mathcal{O}(B_{\frac{\delta}{2}}(\bar u))$ and

$$\limsup_{l\to\infty} I^{\varepsilon_l}(v_{\varepsilon_l}) \le I(\bar u).$$

(4.11) and $v_{\varepsilon_l} \in \mathcal{O}(B_{\frac{\delta}{2}}(\bar u)) = B_{\frac{\delta}{2}}(\mathcal{O}(\bar u))$ imply $I^{\varepsilon_l}(u_{\varepsilon_l}) \le I^{\varepsilon_l}(v_{\varepsilon_l})$, and we obtain

$$\limsup_{l\to\infty} I^{\varepsilon_l}(u_{\varepsilon_l}) \le \limsup_{l\to\infty} I^{\varepsilon_l}(v_{\varepsilon_l}) \le I(\bar u). \tag{4.14}$$

Since $\bar u \in \mathcal{S}_{per}^0(-1,1)$, the right-hand side is finite so that $(I^{\varepsilon_l}(u_{\varepsilon_l}))_{l\in\mathbb{N}}$ is bounded, and by Theorem 4.1 (i) we find a subsequence of $(u_{\varepsilon_l})_{l\in\mathbb{N}}$ (not relabelled) and a $\tilde u \in \mathcal{S}_{per}^0(-1,1)$ such that

$$\lim_{l\to\infty} \|u_{\varepsilon_l} - \tilde u\|_{H^1(-1,1)} = 0. \tag{4.15}$$

Since $(\tau_l)_{l\in\mathbb{N}} \subset [0,2]$, we may assume that $\tau_l \to \tau \in [0,2]$ for $l \to \infty$, and taking the limit for $l \to \infty$ in (4.13) yields

$$\|\tilde u - T_\tau\bar u\|_{H^1(-1,1)} = \frac{\delta}{2},$$

hence $\tilde u \in \overline{B_{\frac{\delta}{2}}(\mathcal{O}(\bar u))} \subset B_\delta(\mathcal{O}(\bar u)) = \mathcal{O}(B_\delta(\bar u))$. Using (4.15) with Theorem 4.1 (ii) and then applying (4.14) we deduce

$$I(\tilde u) \le \liminf_{l\to\infty} I^{\varepsilon_l}(u_{\varepsilon_l}) \le \limsup_{l\to\infty} I^{\varepsilon_l}(u_{\varepsilon_l}) \le I(\bar u).$$

Since $\tilde u \in \mathcal{O}(B_\delta(\bar u))$, we can choose $\sigma \in [0,2]$ such that $T_\sigma\tilde u \in B_\delta(\bar u)$. Then $T_\sigma\tilde u \in \mathcal{S}_{per}^0(-1,1)$, and we obtain

$$I(T_\sigma\tilde u) = I(\tilde u) \le I(\bar u).$$

Using $T_\sigma \tilde{u} \in B_\delta(\bar{u})$ with $\bar{u}$, $T_\sigma \tilde{u} \in \mathcal{S}^0_{per}(-1,1)$, we infer from the assumption in the Theorem that $T_\sigma \tilde{u} \in \mathcal{O}(\bar{u})$ and thus

$$\tilde{u} \in \mathcal{O}(\bar{u}).$$

However, using (4.15), this implies that we find $l_0 \in \mathbb{N}$ such that $u_{\varepsilon_l} \in B_{\frac{\delta}{2}}(\mathcal{O}(\bar{u})) = \mathcal{O}(B_{\frac{\delta}{2}}(\bar{u}))$ for every $l \geq l_0$, which contradicts (4.12). Thus, $u_\varepsilon \in B_{\frac{\delta}{2}}(\mathcal{O}(\bar{u}))$ for $\varepsilon > 0$ small, so that (4.11) implies

$$I^\varepsilon(u_\varepsilon) \leq \inf_{u \in B_{\frac{\delta}{2}}(\mathcal{O}(\bar{u}))} I^\varepsilon(u) \leq I^\varepsilon(u_\varepsilon)$$

for small $\varepsilon > 0$. The inequalities become equalities, and the first part of the Theorem is proved.

To show (4.8), assume for a contradiction that it does not hold true. Then we find a $\delta_0 > 0$ and sequence $(\varepsilon_l)_{l \in \mathbb{N}}$ with $\varepsilon_l \to 0$ for $l \to \infty$ such that

$$\inf_{u \in \mathcal{O}(\bar{u})} \|u_{\varepsilon_l} - u\|_{H^1(-1,1)} \geq \delta_0 \quad \text{for every} \quad l \in \mathbb{N}.$$

Since $u_{\varepsilon_l} \in B_{\frac{\delta}{2}}(\mathcal{O}(\bar{u}))$, for every $l \in \mathbb{N}$ we can choose $\tau_l \in [0,2]$ such that

$$\frac{\delta}{2} > \|u_{\varepsilon_l} - T_{\tau_l}\bar{u}\|_{H^1(-1,1)} \geq \inf_{u \in \mathcal{O}(\bar{u})} \|u_{\varepsilon_l} - u\|_{H^1(-1,1)} \geq \delta_0 > 0 \qquad (4.16)$$

for every $l \in \mathbb{N}$. Again, we use Theorem 4.1(iii) to find a sequence $(v_{\varepsilon_l})_{l \in \mathbb{N}} \subset H^{2,0}_{per}(-1,1)$ such that $(v_{\varepsilon_l}) \in B_{\frac{\delta}{2}}(\bar{u}) \subset \mathcal{O}(B_{\frac{\delta}{2}}(\bar{u}))$ for every $l \in \mathbb{N}$ and

$$\limsup_{l \to \infty} I^{\varepsilon_l}(v_{\varepsilon_l}) \leq I(\bar{u}).$$

As before, (4.11) and $v_{\varepsilon_l} \in \mathcal{O}(B_{\frac{\delta}{2}}(\bar{u})) = B_{\frac{\delta}{2}}(\mathcal{O}(\bar{u}))$ yield $I^{\varepsilon_l}(u_{\varepsilon_l}) \leq I^{\varepsilon_l}(v_{\varepsilon_l})$, so that

$$\limsup_{l \to \infty} I^{\varepsilon_l}(u_{\varepsilon_l}) \leq \limsup_{l \to \infty} I^{\varepsilon_l}(v_{\varepsilon_l}) \leq I(\bar{u})$$

holds. The right-hand is finite due to $\bar{u} \in \mathcal{S}^0_{per}(-1,1)$, so $(I^{\varepsilon_l}(u_{\varepsilon_l}))_{l \in \mathbb{N}}$ is bounded, and by Theorem 4.1(i) we find a $\tilde{u} \in \mathcal{S}^0_{per}(-1,1)$ for which we may assume that

$$\lim_{l \to \infty} \|u_{\varepsilon_l} - \tilde{u}\|_{H^1(-1,1)} = 0, \qquad (4.17)$$

and Theorem 4.1(ii) yields

$$I(\tilde{u}) \leq \liminf_{l \to \infty} I^{\varepsilon_l}(u_{\varepsilon_l}) \leq \limsup_{l \to \infty} I^{\varepsilon_l}(u_{\varepsilon_l}) \leq I(\bar{u}). \qquad (4.18)$$

As before, we choose a subsequence (not relabelled) of $(\tau_l)_{l \in \mathbb{N}}$ such that $\tau_l \to 0$ for $l \to \infty$, and after taking the limit in (4.16), we obtain

$$\frac{\delta}{2} \geq \|\tilde{u} - T_\tau \bar{u}\|_{H^1(-1,1)} = \|T_{-\tau}\tilde{u} - \bar{u}\|_{H^1(-1,1)} \geq \delta_0$$

so that $\tilde{u} \in B_\delta(\mathcal{O}(\bar{u}))$ with $T_{-\tau}\tilde{u} \in B_\delta(\bar{u})$, and from (4.18) we get

$$I(T_{-\tau}\tilde{u}) = I(\tilde{u}) \leq I(\bar{u}).$$

Due to the initial assumption, $T_{-\tau}\tilde{u} \in B_\delta(\bar{u})$ combined with the fact that $T_{-\tau}\tilde{u}$, $\bar{u} \in \mathcal{S}^0_{per}(-1,1)$ implies $T_{-\tau}\tilde{u} \in \mathcal{O}(\bar{u})$, i.e. $\tilde{u} \in \mathcal{O}(\bar{u})$, and (4.16) yields

$$0 < \delta_0 \leq \inf_{u \in \mathcal{O}(\bar{u})} \|u_{\varepsilon_l} - u\|_{H^1(-1,1)} \leq \|u_{\varepsilon_l} - \tilde{u}\|_{H^1(-1,1)},$$

and since the right-hand side converges to zero for $l \to \infty$ (cf. (4.17)), we get a contradiction, and the Theorem follows. □

Now that we have shown Theorem 4.6, it is easy to derive the actual local minimum property:

Corollary 4.7 *Let $\bar{u} \in \mathcal{S}^0_{per}(-1,1)$. Assume that there exists a $\delta > 0$ such that for every $u \in \mathcal{S}^0_{per}(-1,1)$ with $\|u - \bar{u}\|_{H^1(-1,1)} < \delta$ and $u \notin \mathcal{O}(\bar{u})$ we have*

$$I(u) > I(\bar{u}).$$

Then there exists $\varepsilon_0 > 0$ such that for every $\varepsilon \in (0, \varepsilon_0)$ we find $u_\varepsilon \in H^{2,0}_{per}(-1,1)$ with $\|u_\varepsilon - \bar{u}\|_{H^1(-1,1)} < \frac{\delta}{2}$ and

$$I^\varepsilon(u_\varepsilon) \leq I^\varepsilon(u) \quad \text{for every} \quad u \in H^{2,0}_{per}(-1,1) \quad \text{satisfying} \quad \|u - \bar{u}\|_{H^1(-1,1)} < \frac{\delta}{2}.$$

Furthermore,

$$\lim_{\varepsilon \to 0} \|u_\varepsilon - \bar{u}\|_{H^1(-1,1)} = 0.$$

In particular, for every $\varepsilon \in (0, \varepsilon_0)$, u_ε is a local $H^1(-1,1)$-minimizer of I^ε in $H^{2,0}_{per}(-1,1)$.

Proof of Corollary 4.7: Because of Theorem 4.6, there is an $\varepsilon_0 > 0$ such that for every $\varepsilon \in (0, \varepsilon_0)$ we find $v_\varepsilon \in H^{2,0}_{per}(-1,1)$ with $v_\varepsilon \in B_{\frac{\delta}{2}}(\mathcal{O}(\bar{u})) = \mathcal{O}(B_{\frac{\delta}{2}}(\bar{u}))$ and

$$I^\varepsilon(v_\varepsilon) \leq I^\varepsilon(u) \quad \text{for every} \quad u \in B_{\frac{\delta}{2}}(\mathcal{O}(\bar{u})) \cap H^{2,0}_{per}(-1,1),$$

$$\inf_{u \in \mathcal{O}(\bar{u})} \|v_\varepsilon - u\|_{H^1(-1,1)} \to 0 \quad \text{for} \quad \varepsilon \to 0.$$

For every $\varepsilon \in (0, \varepsilon_0)$, we choose $\tau_\varepsilon \in [0,2]$ such that

$$\inf_{u \in \mathcal{O}(\bar{u})} \|v_\varepsilon - u\|_{H^1(-1,1)} \geq \|v_\varepsilon - T_{\tau_\varepsilon}\bar{u}\|_{H^1(-1,1)} - \varepsilon.$$

Setting $u_\varepsilon = T_{-\tau_\varepsilon}v_\varepsilon \in H^{2,0}_{per}(-1,1)$ for every $\varepsilon \in (0, \varepsilon_0)$, we deduce

$$\lim_{\varepsilon \to 0} \|u_\varepsilon - \bar{u}\|_{H^1(-1,1)} = \lim_{\varepsilon \to 0} \|v_\varepsilon - T_{\tau_\varepsilon}\bar{u}\|_{H^1(-1,1)} = 0.$$

In particular, $\|u_\varepsilon - \bar{u}\|_{H^1(-1,1)} < \frac{\delta}{2}$ if ε sufficiently small, and for every $u \in H_{per}^{2,0}(-1,1)$ satisfying $\|u - \bar{u}\|_{H^1(-1,1)} < \frac{\delta}{2}$ we get

$$I^\varepsilon(u_\varepsilon) = I^\varepsilon(T_{\tau_\varepsilon} u_\varepsilon) = I^\varepsilon(v_\varepsilon) \le I^\varepsilon(u).$$

The local minimizing property now easily follows from the fact that for $\varepsilon > 0$ sufficiently small, the set of all $u \in H_{per}^{2,0}(-1,1)$ with $\|u - \bar{u}\|_{H^1(-1,1)} < \frac{\delta}{2}$ is an open environment of u_ε with respect to the $H^1(-1,1)$-topology on $H_{per}^{2,0}(-1,1)$. □

The following theorem states that in order to find local minima of I in the H^1-topology, we can fix a number N of corners in $[-1,1)$ and then look for local minima of E in the set of all periodic sawtooth functions with N corners and zero average. Intuitively, this is easy to see since we can find small perturbations of a sawtooth function u that have more corners than u, which leads to a slight change of the nonlocal energy E, while the counting term in I makes a jump, i.e. is not continuous under small perturbations of u.

Theorem 4.8 *Let* $\bar{u} \in \mathcal{S}_{per}^0(-1,1)$ *with* $\#(S\bar{u}' \cap [-1,1)) = N$. *Then the following two assertions are equivalent:*

(i) *There exists* $\delta > 0$ *such that for every* $u \in \mathcal{S}_{per}^0(-1,1)$ *with* $\|u - \bar{u}\|_{H^1(-1,1)} < \delta$ *and* $u \notin \mathcal{O}(\bar{u})$, *we have*
$$I(\bar{u}) < I(u).$$

(ii) *There exists* $\delta > 0$ *such that for every* $u \in \mathcal{S}_{per}^0(-1,1)$ *with* $\|u - \bar{u}\|_{H^1(-1,1)} < \delta$, $u \notin \mathcal{O}(\bar{u})$ *and* $\#(Su' \cap [-1,1)) = N$, *we have*
$$E(\bar{u}) < E(u).$$

Remark: For $u \in \mathcal{S}_{per}^0(-1,1)$, $N = \#(Su' \cap [-1,1))$ is an even number, since u is periodic: Let the number of corners of u in $(-1,1)$ be odd. Then if, say, $u' = -1$ on $[-1, -1 + \delta_1]$, we have $u' = 1$ on $[1 - \delta_2, 1]$ for some $\delta_2 > 0$, so an additional corner is located at -1. If the number of inner corners is even, the slope of u near -1 and 1 is the same, so -1 defines no additional corner.

Proof of Theorem 4.8: (i)⇒(ii): Let $\delta > 0$ as described. Then for every $u \in \mathcal{S}_{per}^0(-1,1)$ with $\|u - \bar{u}\|_{H^1(-1,1)} < \delta$, $u \notin \mathcal{O}(\bar{u})$ and $\#(Su' \cap [-1,1)) = N$, we have

$$A_0 N + E(\bar{u}) = I(\bar{u}) < I(u) = A_0 N + E(u),$$

thus $E(\bar{u}) < E(u)$, which proves (ii).

(ii)⇒(i): Let $\delta_1 > 0$ such that $E(\bar{u}) < E(u)$ for every $u \in \mathcal{S}_{per}^0(-1,1)$ that satisfies $\|u - \bar{u}\|_{H^1(-1,1)} < \delta_1$, $u \notin \mathcal{O}(\bar{u})$ and $\#(Su' \cap [-1,1)) = N$. Using Lemma 4.2, we can choose $\delta_2 > 0$ such that

$$E(\bar{u}) - E(u) < A_0$$

for every $u \in \mathcal{S}^0_{per}(-1,1)$ with $\|u - \bar{u}\|_{H^1(-1,1)} < \delta_2$. Furthermore, we can find $\delta_3 > 0$ such that

$$\#(Su' \cap [-1,1)) \geq N \quad \text{for every} \quad u \in \mathcal{S}^0_{per}(-1,1) \quad \text{with} \quad \|u - \bar{u}\|_{H^1(-1,1)} < \delta_3.$$

Indeed, if this is not the case, we are able to choose a sequence $(u_n) \subset \mathcal{S}^0_{per}(-1,1)$ with $\#(Su'_n \cap [-1,1)) \leq N - 2$ for every $n \in \mathbb{N}$ (see remark after theorem) which satisfies $\lim_{n\to\infty} \|u_n - \bar{u}\|_{H^1(-1,1)} = 0$. In particular, $\|u'_n - \bar{u}'\|_{L^2(-1,1)} \to 0$, and by choosing a suitable subsequence, we may assume that $\#(Su'_n \cap [-1,1)) = N_0 \leq N - 2$ for every $n \in \mathbb{N}$. Let $t_1^{(n)}, \ldots, t_{N_0}^{(n)} \in [-1,1)$ with

$$-1 \leq t_1^{(n)} < \cdots < t_{N_0}^{(n)} < 1$$

be the corners of u_n. We may assume that

$$u'_n(x) = (-1)^i \quad \text{for} \quad x \in (t_{i-1}^{(n)}, t_i^{(n)}), \quad i = 1, \ldots, N_0 + 1$$

for every $n \in \mathbb{N}$, where $t_0^{(n)} = -1$, $t_{N_0+1}^{(n)} = 1$. We may also assume that for every $i \in 0, \ldots, N_0 + 1$, we have the convergence

$$\lim_{n\to\infty} t_i^{(n)} = t_i \in [-1,1]$$

with

$$-1 = t_0 \leq t_1 \leq \cdots \leq t_{N_0} \leq t_{N_0+1} = 1.$$

This clearly implies that the functions u'_n converge pointwise a.e. to v defined by

$$v(x) = (-1)^i \quad \text{for} \quad x \in (t_{i-1}, t_i), \quad i = 1, \ldots, N_0 + 1,$$

hence $\lim_{n\to\infty} \|u'_n - v\|_{L^2(-1,1)} = 0$ which implies $\bar{u}' = v$. Thus $\bar{u}'$ has $\leq N - 1$ jumps, i.e. $\#(S\bar{u}' \cap [-1,1)) \leq N - 1$ which is a contradiction.

We choose

$$\delta = \min(\delta_1, \delta_2, \delta_3)$$

and take an arbitrary $u \in \mathcal{S}^0_{per}(-1,1)$ with $\|u - \bar{u}\|_{H^1(-1,1)} < \delta$ and $u \notin \mathcal{O}(\bar{u})$. Then $\#(Su' \cap [-1,1)) \geq N$ due to the choice of δ_3. If $\#(Su' \cap [-1,1)) = N$, we obtain

$$I(\bar{u}) = A_0 N + E(\bar{u}) < A_0 \#(Su' \cap [-1,1)) + E(u) = I(u)$$

from how we chose δ_1. If $\#(Su' \cap [-1,1)) \geq N + 1$, we get

$$I(\bar{u}) = A_0 N + E(\bar{u}) < A_0 N + E(u) + A_0 \leq A_0 \#(Su' \cap [-1,1)) + E(u) = I(u)$$

due to our choice of δ_2, and (i) holds in both cases. □

Corollary 4.7 and Theorem 4.8 now immediately imply

Corollary 4.9 *Let $\bar{u} \in \mathcal{S}_{per}^0(-1,1)$ and $N = \#(S\bar{u}' \cap [-1,1))$ (which is an even number). Assume there exists $\delta > 0$ such that for every $u \in \mathcal{S}_{per}^0(-1,1)$ with $\|u - \bar{u}\|_{H^1(-1,1)} < \delta$, $u \notin \mathcal{O}(\bar{u})$ and $\#(Su' \cap [-1,1)) = N$, we have*

$$E(\bar{u}) < E(u).$$

Then there exists $\varepsilon_0 > 0$ such that for every $\varepsilon \in (0, \varepsilon_0)$ we find $u_\varepsilon \in H_{per}^{2,0}(-1,1)$ with $\|u_\varepsilon - \bar{u}\|_{H^1(-1,1)} < \frac{\delta}{2}$ and

$$I^\varepsilon(u_\varepsilon) \leq I^\varepsilon(u) \quad \text{for every} \quad u \in H_{per}^{2,0}(-1,1) \quad \text{satisfying} \quad \|u - \bar{u}\|_{H^1(-1,1)} < \frac{\delta}{2}.$$

Furthermore,

$$\lim_{\varepsilon \to 0} \|u_\varepsilon - \bar{u}\|_{H^1(-1,1)} = 0.$$

In particular, for every $\varepsilon \in (0, \varepsilon_0)$, u_ε is a local $H^1(-1,1)$-minimizer of I^ε in the space $H_{per}^{2,0}(-1,1)$.

Chapter 5

Variation of Sawtooth Functions

5.1   Shifting Corners

Corollary 4.9 suggests that if we find local minimizers of E in the set of all $u \in \mathcal{S}^0_{per}(-1,1)$ with N corners, N a fixed even number, then we can characterize certain local minimizers of I^ε for small $\varepsilon > 0$. For now, we drop the periodicity and zero average conditions and consider the nonlocal energy

$$E(u) = \int_{-1}^{1} \int_{-1}^{1} g(x-y)(u(x) - u(y))^2 dxdy$$

as a functional on the set

$$\mathcal{S}(-1,1) = \{u : [-1,1] \to \mathbb{R} \,|\, u' \in BV((-1,1), \{-1,1\})\}$$

of sawtooth functions with slope $u' \in \{\pm 1\}$ a.e. on $(-1,1)$ and a finite number of corners. For a fixed $u \in \mathcal{S}(-1,1)$, we will define certain functions $F_{u,r} : (-\delta_0, \delta_0) \to \mathbb{R}$ describing the change of energy of this particular sawtooth function under small variations of u that keep the condition $u' \in \{-1,1\}$ without changing the number of corners of u in $(-1,1)$ (in what follows, the number of corners is denoted by N). We will prove that these functions are of class C^2, which is needed later on to specify certain local minimizers of E in the set of all $u \in \mathcal{S}^0_{per}(-1,1)$ with N corners.

Let $u \in \mathcal{S}(-1,1)$ be fixed. Since $E(u) = E(-u)$, we may assume that u starts with slope $+1$, so that we will find numbers

$$-1 = t_0 < t_1 < \cdots < t_{N-1} < t_N = 1 \tag{5.1}$$

such that

$$u'(x) = (-1)^{i-1} \text{ for } x \in (t_{i-1}, t_i) \tag{5.2}$$

for $i = 1, \ldots, N$. Clearly, $t_1, \ldots, t_{N-1}$ are the inner corners of u, or equivalently, the jumps of u' in $(-1,1)$. In order to define a variation of u in the set $\mathcal{S}(-1,1)$ that does not change the number of corners in $(-1,1)$, we fix an arbitrary vector $r = (r_1, \ldots, r_{N-1}) \in \mathbb{R}^{N-1}$

whose components r_i describe the "shifting degree" of the respective inner corners t_i of u. Setting $r_0 = r_N = 0$, for $\delta \in (-\delta_0, \delta_0)$ we define

$$t_i^{(\delta)} = t_i + r_i \delta \quad \text{for } i = 0, \ldots, N, \tag{5.3}$$

where $\delta_0 > 0$ is chosen such that for every $\delta \in (-\delta_0, \delta_0)$, we have

$$-1 = t_0^{(\delta)} < t_1^{(\delta)} < \cdots < t_{N-1}^{(\delta)} < t_N^{(\delta)} = 1 \tag{5.4}$$

which is possible due to (5.1). The condition $r_0 = r_N = 0$ implies $t_0^{(\delta)} = t_0 = -1$ and $t_N^{(\delta)} = t_N = 1$ which means that only inner corners are shifted. Furthermore, no new corners can emerge at the border, which is essential to keep the number of inner corners. For every $\delta \in (-\delta_0, \delta_0)$, we define $u_\delta \in \mathcal{S}(-1, 1)$ as the unique function satisfying

$$\begin{cases} u_\delta(-1) = u(-1) \\ u_\delta'(x) = (-1)^{i-1} \text{ for } x \in (t_{i-1}^{(\delta)}, t_i^{(\delta)}), \quad i = 1, \ldots, N \end{cases} \tag{5.5}$$

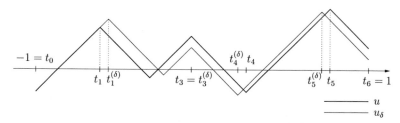

Figure 5.1: Example of a perturbation u_δ of u with $r_3 = 0$, case $N = 6$

and define a function $F_{u,r} : (-\delta_0, \delta_0) \to \mathbb{R}$ by

$$F_{u,r}(\delta) = E(u_\delta) \tag{5.6}$$

for every $\delta \in (-\delta_0, \delta_0)$. Clearly, we have $u_0 = u$, and thus $F_{u,r}(0) = E(u)$.

Remarks:

a) At first glance, the functions u_δ may not cover all variations of the desired type since a function in $\mathcal{S}(-1, 1)$ "close to" u might not satisfy $u_\delta(-1) = u(-1)$. Also, in the case of a periodic function u which has a corner at -1, above construction shift inner corners only, thus -1 is fixed as a corner. These defects, however, are adjusted by the invariance of E under translations and the addition of constants. For the moment, I just point out that this method is sufficient for what we want to prove.

b) We omit the dependence of u_δ, $t_i^{(\delta)}$ on r to avoid ugly notation but will indicate it in Chapter 8, where it becomes necessary. For now, we consider a fixed r.

5.2 Marginal Change of Nonlocal Energy

The following results show that the functions $F_{u,r}$ are of class C^2 on $(-\delta_0, \delta_0)$. For our calculations it is convenient to introduce a symmetric bilinear form defined by

$$(u, v)_g = \int_{-1}^{1} \int_{-1}^{1} g(x - y)(u(x) - u(y))(v(x) - v(y)) dx dy \qquad (5.7)$$

which due to Lebesgue's dominated convergence theorem is continuous under pointwise convergence of sequences u_n, v_n with $|u_n(x)|$, $|v_n(x)| \leq C$ a.e. in $[-1, 1]$ for every $n \in \mathbb{N}$. Furthermore, we have $E(u) = (u, u)_g$. To compute the derivatives of $F_{u,r}$, we need the following technical Lemma:

Lemma 5.1 *With the above definitions, let $\delta \in (-\delta_0, \delta_0)$. Then*

(i) For a.e. $x \in (-1, 1)$, we have

$$\lim_{\varepsilon \to 0} \frac{1}{\varepsilon}(u_{\delta+\varepsilon}(x) - u_\delta(x)) = 2 \sum_{i=1}^{N} \sum_{j=1}^{i-1} (-1)^{j-1} r_j \chi_{(t_{i-1}^{(\delta)}, t_i^{(\delta)})} \qquad (5.8)$$

$$\left| \frac{1}{\varepsilon}(u_{\delta+\varepsilon}(x) - u_\delta(x)) \right| \leq C$$

with C independent of δ, ε, x (but dependent on r).

(ii) For a.e. $x \in (-1, 1)$, we have

$$\lim_{\varepsilon \to 0}(u_{\delta+\varepsilon}(x) - u_\delta(x)) = 0$$

and $|u_\delta(x)|$, $|u_{\delta+\varepsilon}(x)| \leq C$, with C independent of δ, ε, x (but dependent on u).

Proof: For fixed $\delta \in (-\delta_0, \delta_0)$, choose $\varepsilon_0 > 0$ such that $(\delta - \varepsilon_0, \delta + \varepsilon_0) \subset (-\delta_0, \delta_0)$ and $t_{i-1}^{(\delta+\varepsilon)} \vee t_{i-1}^{(\delta)} < t_i^{(\delta+\varepsilon)} \wedge t_i^{(\delta)}$ for every $\varepsilon \in (-\varepsilon_0, \varepsilon_0)$ and $i \in \{1, \ldots, N\}$ (which is possible because of (5.4) and since $t_i^{(\delta+\varepsilon)} \to t_i^{(\delta)}$ for $\varepsilon \to 0$ for every $i \in \{0, \ldots, N\}$). Then, according to (5.5), we have

$$(u'_{\delta+\varepsilon} - u'_\delta)(x) = 0 \text{ for } x \in (t_{i-1}^{(\delta)}, t_i^{(\delta)}) \cap (t_{i-1}^{(\delta+\varepsilon)}, t_i^{(\delta+\varepsilon)}) = (t_{i-1}^{(\delta+\varepsilon)} \vee t_{i-1}^{(\delta)}, t_i^{(\delta+\varepsilon)} \wedge t_i^{(\delta)})$$

for $i = 1, \ldots, N$. Thus, up to a finite subset of $[-1, 1]$, $(u'_{\delta+\varepsilon} - u'_\delta)(x) \neq 0$ can only hold if $x \in (t_i^{(\delta+\varepsilon)} \wedge t_i^{(\delta)}, t_i^{(\delta+\varepsilon)} \vee t_i^{(\delta+\varepsilon)})$ for an $i \in \{1, \ldots, N-1\}$. Consider a fixed $x \in (t_{i-1}^{(\delta)}, t_i^{(\delta)})$, $i \in \{1, \ldots, N\}$. Then for small $\varepsilon \in (-\varepsilon_0, \varepsilon_0)$, we have $x \in (t_{i-1}^{(\delta+\varepsilon)} \vee t_{i-1}^{(\delta)}, t_i^{(\delta+\varepsilon)} \wedge t_i^{(\delta)})$ since $t_{i-1}^{(\delta+\varepsilon)} \to t_{i-1}^{(\delta)}$ and $t_i^{(\delta+\varepsilon)} \to t_i^{(\delta)}$ for $\varepsilon \to 0$. Combining these facts we get, using $u_\delta(-1) = u_{\delta+\varepsilon}(-1)$,

$$u_{\delta+\varepsilon}(x) - u_\delta(x) = \int_{-1}^{x} (u'_{\delta+\varepsilon} - u'_\delta)(\xi) d\xi = \sum_{j=1}^{i-1} \int_{t_j^{(\delta+\varepsilon)} \wedge t_j^{(\delta)}}^{t_j^{(\delta+\varepsilon)} \vee t_j^{(\delta)}} (u'_{\delta+\varepsilon} - u'_\delta)(\xi) d\xi \qquad (5.9)$$

for $x \in (t_{i-1}^{(\delta)}, t_i^{(\delta)})$ if ε is sufficiently small. Let $j \in \{1, \ldots, N\}$. Then if $r_j \varepsilon > 0$, we have $t_j^{(\delta+\varepsilon)} > t_j^{(\delta)}$, and we get

$$\int_{t_j^{(\delta+\varepsilon)} \wedge t_j^{(\delta)}}^{t_j^{(\delta+\varepsilon)} \vee t_j^{(\delta)}} (u_{\delta+\varepsilon}' - u_\delta')(\xi) d\xi = \int_{t_j^{(\delta)}}^{t_j^{(\delta+\varepsilon)}} (u_{\delta+\varepsilon}' - u_\delta')(\xi) d\xi = 2(-1)^{j-1} r_j \varepsilon$$

for ε sufficiently small since

$$\begin{aligned} u_\delta' &= (-1)^j & \text{on} \quad (t_j^{(\delta)}, t_j^{(\delta+\varepsilon)}) \subset (t_j^{(\delta)}, t_{j+1}^{(\delta)}) & \quad \text{and} \\ u_{\delta+\varepsilon}' &= (-1)^{j-1} & \text{on} \quad (t_j^{(\delta)}, t_j^{(\delta+\varepsilon)}) \subset (t_{j-1}^{(\delta+\varepsilon)}, t_j^{(\delta+\varepsilon)}). \end{aligned}$$

If $r_j \varepsilon < 0$, we have $t_j^{(\delta+\varepsilon)} < t_j^{(\delta)}$, and we deduce

$$\int_{t_j^{(\delta+\varepsilon)} \wedge t_j^{(\delta)}}^{t_j^{(\delta+\varepsilon)} \vee t_j^{(\delta)}} (u_{\delta+\varepsilon}' - u_\delta')(\xi) d\xi = \int_{t_j^{(\delta+\varepsilon)}}^{t_j^{(\delta)}} (u_{\delta+\varepsilon}' - u_\delta')(\xi) d\xi = 2(-1)^j (-r_j \varepsilon) = 2(-1)^{j-1} r_j \varepsilon$$

for sufficiently small ε from

$$\begin{aligned} u_\delta' &= (-1)^{j-1} & \text{on} \quad (t_j^{(\delta+\varepsilon)}, t_j^{(\delta)}) \subset (t_{j-1}^{(\delta)}, t_j^{(\delta)}) & \quad \text{and} \\ u_{\delta+\varepsilon}' &= (-1)^j & \text{on} \quad (t_j^{(\delta+\varepsilon)}, t_j^{(\delta)}) \subset (t_j^{(\delta+\varepsilon)}, t_{j+1}^{(\delta+\varepsilon)}). \end{aligned}$$

If $r_j \varepsilon = 0$, then $t_j^{(\delta)} = t_j^{(\delta+\varepsilon)}$ (cf. (5.3)), i.e. $t_j^{(\delta)} \wedge t_j^{(\delta+\varepsilon)} = t_j^{(\delta)} \vee t_j^{(\delta+\varepsilon)}$ and thus we get the same formula as both sides are zero. Setting this into (5.9), we obtain

$$u_{\delta+\varepsilon}(x) - u_\delta(x) = 2\varepsilon \sum_{j=1}^{i-1} (-1)^{j-1} r_j \quad \text{for} \quad x \in (t_{i-1}^{(\delta)}, t_i^{(\delta)}), \quad i = 1, \ldots, N$$

if ε is small, and thus

$$u_{\delta+\varepsilon} - u_\delta = 2\varepsilon \sum_{i=1}^N \sum_{j=1}^{i-1} (-1)^{j-1} r_j \chi_{(t_{i-1}^{(\delta)}, t_i^{(\delta)})}$$

for a.e. $x \in (-1, 1)$ if ε is sufficiently small, and we deduce

$$\lim_{\varepsilon \to 0} \frac{1}{\varepsilon} (u_{\delta+\varepsilon} - u_\delta) = 2 \sum_{i=1}^N \sum_{j=1}^{i-1} (-1)^{j-1} r_j \chi_{(t_{i-1}^{(\delta)}, t_i^{(\delta)})}$$

a.e. in $(-1, 1)$. To show the bound in (i), we derive from (5.9) and (5.3)

$$\left| \frac{1}{\varepsilon} (u_{\delta+\varepsilon}(x) - u_\delta(x)) \right| \leq \frac{1}{|\varepsilon|} \sum_{j=1}^{i-1} \int_{t_j^{(\delta+\varepsilon)} \wedge t_j^{(\delta)}}^{t_j^{(\delta+\varepsilon)} \vee t_j^{(\delta)}} |u_{\delta+\varepsilon}'(\xi) - u_\delta'(\xi)| d\xi \leq \frac{2}{|\varepsilon|} \sum_{j=1}^{i-1} |r_i \varepsilon| \leq 2 \sum_{j=1}^{N-1} |r_i|$$

a.e. in $(-1, 1)$. The convergence in (ii) immediately follows from (i), while the bound follows simply from the definition of u_δ:

$$|u_\delta(x)| = \left| \int_{-1}^x u_\delta'(\xi) d\xi + u(-1) \right| \leq 2 + |u(-1)|.$$

$\square$

The derivative of $F_{u,r}$ is characterized in the following Theorem.

Theorem 5.2 *Let $u \in \mathcal{S}(-1,1)$, $r = (r_1, \ldots, r_{N-1}) \in \mathbb{R}^{N-1}$ be arbitrary, $\delta_0 > 0$ chosen as above. Then the function $F_{u,r} : (-\delta_0, \delta_0) \to \mathbb{R}$ as defined in (5.6) is differentiable, and for every $\delta \in (-\delta_0, \delta_0)$, we have*

$$F'_{u,r}(\delta) = 4 \sum_{i=1}^{N} \alpha_i h_{u,r}^{(i)}(\delta) \,, \tag{5.10}$$

where $\alpha_i \in \mathbb{R}$, $h_{u,r}^{(i)} : (-\delta_0, \delta_0) \to \mathbb{R}$ are given by

$$\alpha_i = \sum_{j=1}^{i-1} (-1)^{j-1} r_j \,, \quad h_{u,r}^{(i)}(\delta) = \left(\chi_{(t_{i-1}^{(\delta)}, t_i^{(\delta)})}, u_\delta \right)_g \tag{5.11}$$

for $i = 1, \ldots, N$.

Proof: Let $\delta \in (-\delta_0, \delta_0)$, $\varepsilon_0 > 0$ such that $(\delta - \varepsilon_0, \delta + \varepsilon_0) \subset (-\delta_0, \delta_0)$. Then for every $\varepsilon \in (-\varepsilon_0, \varepsilon_0)$

$$F_{u,r}(\delta + \varepsilon) - F_{u,r}(\delta) = (u_{\delta+\varepsilon}, u_{\delta+\varepsilon})_g - (u_\delta, u_\delta)_g = (u_{\delta+\varepsilon} - u_\delta, u_{\delta+\varepsilon} + u_\delta)_g \,. \tag{5.12}$$

Hence

$$\frac{1}{\varepsilon}(F_{u,r}(\delta + \varepsilon) - F_{u,r}(\delta)) = \left(\frac{1}{\varepsilon}(u_{\delta+\varepsilon} - u_\delta), u_{\delta+\varepsilon} + u_\delta \right)_g \,,$$

which due to Lemma 5.1 and Lebesgue's convergence theorem, yields

$$\lim_{\varepsilon \to 0} \frac{1}{\varepsilon}(F_{u,r}(\delta + \varepsilon) - F_{u,r}(\delta)) = 4 \sum_{i=1}^{N} \sum_{j=1}^{i-1} (-1)^{j-1} r_j \left(\chi_{(t_{i-1}^{(\delta)}, t_i^{(\delta)})}, u_\delta \right)_g \,,$$

i.e.

$$F'_{u,r}(\delta) = \lim_{\varepsilon \to 0} \frac{1}{\varepsilon}(F_{u,r}(\delta + \varepsilon) - F_{u,r}(\delta)) = 4 \sum_{i=1}^{N} \alpha_i h_{u,r}^{(i)}(\delta) \,,$$

with $\alpha_i \in \mathbb{R}$, $h_{u,r}^{(i)} : (-\delta_0, \delta_0) \to \mathbb{R}$, $i = 1, \ldots, N$, as given in (5.11), which completes the proof of Theorem 5.2. $\qquad\qquad \square$

5.3 The Second Derivative

The following theorem deals with the second derivative of $F_{u,r}$:

Theorem 5.3 *For $u \in \mathcal{S}(-1,1)$, $r = (r_1, \ldots, r_{N-1}) \in \mathbb{R}^{N-1}$, choose $\delta_0 > 0$ as above. Then the function $F_{u,r} : (-\delta_0, \delta_0) \to \mathbb{R}$ as defined in (5.6) satisfies $F_{u,r} \in C^2(-\delta_0, \delta_0)$, and for every $\delta \in (-\delta_0, \delta_0)$, we have*

$$F''_{u,r}(\delta) = 8 \sum_{i=1}^{N} \sum_{k=1}^{N} \alpha_i \alpha_k \left(\chi_{(t_{k-1}^{(\delta)}, t_k^{(\delta)})}, \chi_{(t_{i-1}^{(\delta)}, t_i^{(\delta)})} \right)_g$$

$$+8\sum_{i=1}^{N}\alpha_i r_i \int_{-1}^{1} g(t_i^{(\delta)}-y)(u_\delta(t_i^{(\delta)})-u_\delta(y))dy \qquad (5.13)$$

$$-8\sum_{i=1}^{N}\alpha_i r_{i-1} \int_{-1}^{1} g(t_{i-1}^{(\delta)}-y)(u_\delta(t_{i-1}^{(\delta)})-u_\delta(y))dy\,,$$

with α_i, $i = 1,\ldots,N$ defined as in (5.11).

Due to Theorem 5.2, this is an immediate consequence of the following:

Theorem 5.4 For $u \in \mathcal{S}(-1,1)$, $r = (r_1,\ldots,r_{N-1}) \in \mathbb{R}^{N-1}$, choose $\delta_0 > 0$ as above, and let $h_{u,r}^{(i)} : (-\delta_0,\delta_0) \to \mathbb{R}$, $i = 1,\ldots,N$ be given as in (5.11). Then $h_{u,r}^{(i)} \in C^1(-\delta_0,\delta_0)$ with

$$\begin{aligned}
(h_{u,r}^{(i)})'(\delta) &= 2\sum_{k=1}^{N}\alpha_k \left(\chi_{(t_{k-1}^{(\delta)},t_k^{(\delta)})},\chi_{(t_{i-1}^{(\delta)},t_i^{(\delta)})}\right)_g \\
&\quad +2r_i \int_{-1}^{1} g(t_i^{(\delta)}-y)(u_\delta(t_i^{(\delta)})-u_\delta(y))dy \qquad (5.14) \\
&\quad -2r_{i-1} \int_{-1}^{1} g(t_{i-1}^{(\delta)}-y)(u_\delta(t_{i-1}^{(\delta)})-u_\delta(y))dy
\end{aligned}$$

for every $\delta \in (-\delta_0,\delta_0)$, $i = 1,\ldots,N$, where α_i, $i = 1,\ldots,N$ are defined as in (5.11).

Proof: Fix $\delta \in (-\delta_0,\delta_0)$. As in the proof of Lemma 5.1, we choose $\varepsilon_0 > 0$ such that $(\delta - \varepsilon_0,\delta + \varepsilon_0) \subset (-\delta_0,\delta_0)$ and $t_{i-1}^{(\delta+\varepsilon)} \vee t_{i-1}^{(\delta)} < t_i^{(\delta+\varepsilon)} \wedge t_i^{(\delta)}$ for every $\varepsilon \in (-\varepsilon_0,\varepsilon_0)$ and $i \in \{1,\ldots,N\}$. Let $h_{u,r}^{(i)} : (-\delta_0,\delta_0) \to \mathbb{R}$ as given in (5.11), then for $\varepsilon \in (-\varepsilon_0,\varepsilon_0)$

$$h_{u,r}^{(i)}(\delta+\varepsilon)-h_{u,r}^{(i)}(\delta) = \left(\chi_{(t_{i-1}^{(\delta+\varepsilon)},t_i^{(\delta+\varepsilon)})} - \chi_{(t_{i-1}^{(\delta)},t_i^{(\delta)})},u_\delta\right)_g + \left(u_{\delta+\varepsilon} - u_\delta,\chi_{(t_{i-1}^{(\delta+\varepsilon)},t_i^{(\delta+\varepsilon)})}\right)_g \quad (5.15)$$

and using Lemma 5.1(i), Lebesgue's convergence theorem and (5.11), we deduce

$$\begin{aligned}
\lim_{\varepsilon\to 0}\frac{1}{\varepsilon}\left(u_{\delta+\varepsilon} - u_\delta,\chi_{(t_{i-1}^{(\delta+\varepsilon)},t_i^{(\delta+\varepsilon)})}\right)_g &= 2\sum_{k=1}^{N}\sum_{l=1}^{k-1}(-1)^{l-1}r_l \left(\chi_{(t_{k-1}^{(\delta)},t_k^{(\delta)})},\chi_{(t_{i-1}^{(\delta)},t_i^{(\delta)})}\right)_g \\
&= 2\sum_{k=1}^{N}\alpha_k \left(\chi_{(t_{k-1}^{(\delta)},t_k^{(\delta)})},\chi_{(t_{i-1}^{(\delta)},t_i^{(\delta)})}\right)_g. \qquad (5.16)
\end{aligned}$$

Computing the other term in the right-hand side of (5.15), however, is more difficult, and one has to consider several cases:

Case 1: $r_{i-1}\varepsilon > 0$, $r_i\varepsilon > 0$. Then $t_{i-1}^{(\delta)} < t_{i-1}^{(\delta+\varepsilon)} < t_i^{(\delta)} < t_i^{(\delta+\varepsilon)}$ for $\varepsilon \in (-\varepsilon_0,\varepsilon_0)$ due to the choice of ε_0 which yields

$$\chi_{(t_{i-1}^{(\delta+\varepsilon)},t_i^{(\delta+\varepsilon)})} - \chi_{(t_{i-1}^{(\delta)},t_i^{(\delta)})} = -\chi_{(t_{i-1}^{(\delta)},t_{i-1}^{(\delta+\varepsilon)})} + \chi_{(t_i^{(\delta)},t_i^{(\delta+\varepsilon)})}$$

a.e. in $(-1, 1)$, and we get

$$\left(\chi_{(t_{i-1}^{(\delta+\varepsilon)}, t_i^{(\delta+\varepsilon)})} - \chi_{(t_{i-1}^{(\delta)}, t_i^{(\delta)})}, u_\delta\right)_g = \int_{-1}^{1}\int_{-1}^{1} g(x-y)(u_\delta(x) - u_\delta(y))\phi_{\delta,\varepsilon}(x,y)dydx\,,$$

where

$$\phi_{\delta,\varepsilon}(x,y) = -\chi_{(t_{i-1}^{(\delta)}, t_{i-1}^{(\delta+\varepsilon)})}(x) + \chi_{(t_i^{(\delta)}, t_i^{(\delta+\varepsilon)})}(x) + \chi_{(t_{i-1}^{(\delta)}, t_{i-1}^{(\delta+\varepsilon)})}(y) - \chi_{(t_i^{(\delta)}, t_i^{(\delta+\varepsilon)})}(y)\,.$$

Keeping in mind that $(t_{i-1}^{(\delta)}, t_{i-1}^{(\delta+\varepsilon)})$ and $(t_i^{(\delta)}, t_i^{(\delta+\varepsilon)})$ are disjoint due to the choice of ε_0, we see that $\phi_{\delta,\varepsilon}$ vanishes on $S_1 \cup S_2 \cup S_3$, where

$$
\begin{aligned}
S_1 &= (t_{i-1}^{(\delta)}, t_{i-1}^{(\delta+\varepsilon)}) \times (t_{i-1}^{(\delta)}, t_{i-1}^{(\delta+\varepsilon)}) \\
S_2 &= (t_i^{(\delta)}, t_i^{(\delta+\varepsilon)}) \times (t_i^{(\delta)}, t_i^{(\delta+\varepsilon)}) \\
S_3 &= ((-1, t_{i-1}^{(\delta)}) \cup (t_{i-1}^{(\delta+\varepsilon)}, t_i^{(\delta)}) \cup (t_i^{(\delta+\varepsilon)}, 1)) \times ((-1, t_{i-1}^{(\delta)}) \cup (t_{i-1}^{(\delta+\varepsilon)}, t_i^{(\delta)}) \cup (t_i^{(\delta+\varepsilon)}, 1))\,,
\end{aligned}
$$

so that we just have to integrate over $([-1,1] \times [-1,1]) \setminus (S_1 \cup S_2 \cup S_3)$, which up to a set of empty Lebesgue measure equals $\Omega_1 \cup \Omega_2 \cup \Omega_3 \cup \Omega_4 \cup \Omega_5 \cup \Omega_6$ with

$$
\begin{aligned}
\Omega_1 &= (t_{i-1}^{(\delta)}, t_{i-1}^{(\delta+\varepsilon)}) \times (t_i^{(\delta)}, t_i^{(\delta+\varepsilon)}) \\
\Omega_2 &= (t_i^{(\delta)}, t_i^{(\delta+\varepsilon)}) \times (t_{i-1}^{(\delta)}, t_{i-1}^{(\delta+\varepsilon)}) \\
\Omega_3 &= (t_{i-1}^{(\delta)}, t_{i-1}^{(\delta+\varepsilon)}) \times ((-1, t_{i-1}^{(\delta)}) \cup (t_{i-1}^{(\delta+\varepsilon)}, t_i^{(\delta)}) \cup (t_i^{(\delta+\varepsilon)}, 1)) \\
\Omega_4 &= ((-1, t_{i-1}^{(\delta)}) \cup (t_{i-1}^{(\delta+\varepsilon)}, t_i^{(\delta)}) \cup (t_i^{(\delta+\varepsilon)}, 1)) \times (t_{i-1}^{(\delta)}, t_{i-1}^{(\delta+\varepsilon)}) \\
\Omega_5 &= (t_i^{(\delta)}, t_i^{(\delta+\varepsilon)}) \times ((-1, t_{i-1}^{(\delta)}) \cup (t_{i-1}^{(\delta+\varepsilon)}, t_i^{(\delta)}) \cup (t_i^{(\delta+\varepsilon)}, 1)) \\
\Omega_6 &= ((-1, t_{i-1}^{(\delta)}) \cup (t_{i-1}^{(\delta+\varepsilon)}, t_i^{(\delta)}) \cup (t_i^{(\delta+\varepsilon)}, 1)) \times (t_i^{(\delta)}, t_i^{(\delta+\varepsilon)})\,.
\end{aligned}
$$

Using the fact that Ω_i, $i = 1, \ldots, 6$ are pairwise disjoint and the symmetry

$$g(x-y)(u_\delta(x) - u_\delta(y))\phi_{\delta,\varepsilon}(x,y) = g(y-x)(u_\delta(y) - u_\delta(x))\phi_{\delta,\varepsilon}(y,x)\,,$$

which easily follows from the symmetry $g(-t) = g(t)$, we obtain

$$
\begin{aligned}
\left(\chi_{(t_{i-1}^{(\delta+\varepsilon)}, t_i^{(\delta+\varepsilon)})} - \chi_{(t_{i-1}^{(\delta)}, t_i^{(\delta)})}, u_\delta\right)_g &= 2\int_{\Omega_1} g(x-y)(u_\delta(x) - u_\delta(y))\phi_{\delta,\varepsilon}(x,y)dydx \\
&\quad + 2\int_{\Omega_3} g(x-y)(u_\delta(x) - u_\delta(y))\phi_{\delta,\varepsilon}(x,y)dydx \\
&\quad + 2\int_{\Omega_5} g(x-y)(u_\delta(x) - u_\delta(y))\phi_{\delta,\varepsilon}(x,y)dydx \\
&= -4\int_{\Omega_1} g(x-y)(u_\delta(x) - u_\delta(y))dydx \\
&\quad - 2\int_{\Omega_3} g(x-y)(u_\delta(x) - u_\delta(y))dydx \\
&\quad + 2\int_{\Omega_5} g(x-y)(u_\delta(x) - u_\delta(y))dydx \\
&=: I_{1,\varepsilon} + I_{2,\varepsilon} + I_{3,\varepsilon}\,.
\end{aligned}
$$

As for the first line, we compute, using the boundedness of g and u_δ (cf. Lemma 5.1)

$$|I_{1,\varepsilon}| = \left| 4 \int_{t_{i-1}^{(\delta)}}^{t_{i-1}^{(\delta+\varepsilon)}} \int_{t_i^{(\delta)}}^{t_i^{(\delta+\varepsilon)}} g(x-y)(u_\delta(x)-u_\delta(y))dydx \right| \leq C(r_{i-1}\varepsilon)(r_i\varepsilon) \leq C\varepsilon^2 ,$$

i.e. $I_{1,\varepsilon} = O(\varepsilon^2)$. Now examine the second line:

$$I_{2,\varepsilon} = -2 \int_{t_{i-1}^{(\delta)}}^{t_{i-1}^{(\delta+\varepsilon)}} \int_{-1}^{1} g(x-y)(u_\delta(x)-u_\delta(y))dydx$$

$$+ 2 \int_{t_{i-1}^{(\delta)}}^{t_{i-1}^{(\delta+\varepsilon)}} \int_{t_{i-1}^{(\delta)}}^{t_{i-1}^{(\delta+\varepsilon)}} g(x-y)(u_\delta(x)-u_\delta(y))dydx$$

$$+ 2 \int_{t_{i-1}^{(\delta)}}^{t_{i-1}^{(\delta+\varepsilon)}} \int_{t_i^{(\delta)}}^{t_i^{(\delta+\varepsilon)}} g(x-y)(u_\delta(x)-u_\delta(y))dydx$$

$$=: J_{1,\varepsilon} + J_{2,\varepsilon} + J_{3,\varepsilon} .$$

Analogously to $I_{1,\varepsilon}$, we deduce

$$J_{2,\varepsilon} + J_{3,\varepsilon} = O(\varepsilon^2) .$$

Thus, using $t_{i-1}^{(\delta+\varepsilon)} > t_{i-1}^{(\delta)}$ and $r_{i-1}\varepsilon > 0$,

$$I_{2,\varepsilon} = -2\,\mathrm{sgn}(r_{i-1}\varepsilon) \int_{t_{i-1}^{(\delta)} \wedge t_{i-1}^{(\delta+\varepsilon)}}^{t_{i-1}^{(\delta)} \vee t_{i-1}^{(\delta+\varepsilon)}} \int_{-1}^{1} g(x-y)(u_\delta(x)-u_\delta(y))dydx + O(\varepsilon^2) .$$

Arguing as before we rewrite the third line as

$$I_{3,\varepsilon} = 2 \int_{t_i^{(\delta)}}^{t_i^{(\delta+\varepsilon)}} \int_{-1}^{1} g(x-y)(u_\delta(x)-u_\delta(y))dydx$$

$$- 2 \int_{t_i^{(\delta)}}^{t_i^{(\delta+\varepsilon)}} \int_{t_{i-1}^{(\delta)}}^{t_{i-1}^{(\delta+\varepsilon)}} g(x-y)(u_\delta(x)-u_\delta(y))dydx$$

$$- 2 \int_{t_i^{(\delta)}}^{t_i^{(\delta+\varepsilon)}} \int_{t_i^{(\delta)}}^{t_i^{(\delta+\varepsilon)}} g(x-y)(u_\delta(x)-u_\delta(y))dydx$$

$$= 2\,\mathrm{sgn}(r_i\varepsilon) \int_{t_i^{(\delta)} \wedge t_i^{(\delta+\varepsilon)}}^{t_i^{(\delta)} \vee t_i^{(\delta+\varepsilon)}} \int_{-1}^{1} g(x-y)(u_\delta(x)-u_\delta(y))dydx + O(\varepsilon^2) ,$$

and combining all these results we get

$$\left(\chi_{(t_{i-1}^{(\delta+\varepsilon)}, t_i^{(\delta+\varepsilon)})} - \chi_{(t_{i-1}^{(\delta)}, t_i^{(\delta)})}, u_\delta \right)_g =$$

$$= -2\,\mathrm{sgn}(r_{i-1}\varepsilon) \int_{t_{i-1}^{(\delta)} \wedge t_{i-1}^{(\delta+\varepsilon)}}^{t_{i-1}^{(\delta)} \vee t_{i-1}^{(\delta+\varepsilon)}} \int_{-1}^{1} g(x-y)(u_\delta(x)-u_\delta(y))dydx$$

$$+ 2\operatorname{sgn}(r_i\varepsilon) \int_{t_i^{(\delta)} \wedge t_i^{(\delta+\varepsilon)}}^{t_i^{(\delta)} \vee t_i^{(\delta+\varepsilon)}} \int_{-1}^{1} g(x-y)(u_\delta(x) - u_\delta(y))dydx + O(\varepsilon^2).$$

Case 2: $r_{i-1}\varepsilon > 0$, $r_i\varepsilon \leq 0$. Then $t_{i-1}^{(\delta)} < t_{i-1}^{(\delta+\varepsilon)} < t_i^{(\delta+\varepsilon)} \leq t_i^{(\delta)}$ for $\varepsilon \in (-\varepsilon_0, \varepsilon_0)$ and thus

$$\chi_{(t_{i-1}^{(\delta+\varepsilon)}, t_i^{(\delta+\varepsilon)})} - \chi_{(t_{i-1}^{(\delta)}, t_i^{(\delta)})} = -\chi_{(t_{i-1}^{(\delta)}, t_{i-1}^{(\delta+\varepsilon)})} - \chi_{(t_i^{(\delta+\varepsilon)}, t_i^{(\delta)})}$$

a.e. in $(-1,1)$, which yields

$$\left(\chi_{(t_{i-1}^{(\delta+\varepsilon)}, t_i^{(\delta+\varepsilon)})} - \chi_{(t_{i-1}^{(\delta)}, t_i^{(\delta)})}, u_\delta\right)_g = \int_{-1}^{1} \int_{-1}^{1} g(x-y)(u_\delta(x) - u_\delta(y))\phi_{\delta,\varepsilon}(x,y)dydx\,,$$

where

$$\phi_{\delta,\varepsilon}(x,y) = -\chi_{(t_{i-1}^{(\delta)}, t_{i-1}^{(\delta+\varepsilon)})}(x) - \chi_{(t_i^{(\delta+\varepsilon)}, t_i^{(\delta)})}(x) + \chi_{(t_{i-1}^{(\delta)}, t_{i-1}^{(\delta+\varepsilon)})}(y) + \chi_{(t_i^{(\delta+\varepsilon)}, t_i^{(\delta)})}(y).$$

Since $\phi_{\delta,\varepsilon}$ vanishes on $S_1 \cup S_2 \cup S_3 \cup S_4 \cup S_5$, where

$$
\begin{aligned}
S_1 &= (t_{i-1}^{(\delta)}, t_{i-1}^{(\delta+\varepsilon)}) \times (t_{i-1}^{(\delta)}, t_{i-1}^{(\delta+\varepsilon)})\\
S_2 &= (t_{i-1}^{(\delta)}, t_{i-1}^{(\delta+\varepsilon)}) \times (t_i^{(\delta+\varepsilon)}, t_i^{(\delta)})\\
S_3 &= (t_i^{(\delta+\varepsilon)}, t_i^{(\delta)}) \times (t_{i-1}^{(\delta)}, t_{i-1}^{(\delta+\varepsilon)})\\
S_4 &= (t_i^{(\delta+\varepsilon)}, t_i^{(\delta)}) \times (t_i^{(\delta+\varepsilon)}, t_i^{(\delta)})\\
S_5 &= ((-1, t_{i-1}^{(\delta)}) \cup (t_{i-1}^{(\delta+\varepsilon)}, t_i^{(\delta+\varepsilon)}) \cup (t_i^{(\delta)}, 1)) \times ((-1, t_{i-1}^{(\delta)}) \cup (t_{i-1}^{(\delta+\varepsilon)}, t_i^{(\delta+\varepsilon)}) \cup (t_i^{(\delta)}, 1)),
\end{aligned}
$$

we only have to integrate over the following domains

$$
\begin{aligned}
\Omega_1 &= (t_{i-1}^{(\delta)}, t_{i-1}^{(\delta+\varepsilon)}) \times ((-1, t_{i-1}^{(\delta)}) \cup (t_{i-1}^{(\delta+\varepsilon)}, t_i^{(\delta+\varepsilon)}) \cup (t_i^{(\delta)}, 1))\\
\Omega_2 &= (t_i^{(\delta+\varepsilon)}, t_i^{(\delta)}) \times ((-1, t_{i-1}^{(\delta)}) \cup (t_{i-1}^{(\delta+\varepsilon)}, t_i^{(\delta+\varepsilon)}) \cup (t_i^{(\delta)}, 1))\\
\Omega_3 &= ((-1, t_{i-1}^{(\delta)}) \cup (t_{i-1}^{(\delta+\varepsilon)}, t_i^{(\delta+\varepsilon)}) \cup (t_i^{(\delta)}, 1)) \times (t_{i-1}^{(\delta)}, t_{i-1}^{(\delta+\varepsilon)})\\
\Omega_4 &= ((-1, t_{i-1}^{(\delta)}) \cup (t_{i-1}^{(\delta+\varepsilon)}, t_i^{(\delta+\varepsilon)}) \cup (t_i^{(\delta)}, 1)) \times (t_i^{(\delta+\varepsilon)}, t_i^{(\delta)}).
\end{aligned}
$$

These clearly are pairwise disjoint, and as in case 1, due to the symmetry of the integrand, it is enough to integrate over Ω_1 and Ω_2, taking each integral twice. Hence we obtain

$$
\begin{aligned}
\left(\chi_{(t_{i-1}^{(\delta+\varepsilon)}, t_i^{(\delta+\varepsilon)})} - \chi_{(t_{i-1}^{(\delta)}, t_i^{(\delta)})}, u_\delta\right)_g &= 2\int_{\Omega_1} g(x-y)(u_\delta(x) - u_\delta(y))\phi_{\delta,\varepsilon}(x,y)dydx\\
&\quad + 2\int_{\Omega_2} g(x-y)(u_\delta(x) - u_\delta(y))\phi_{\delta,\varepsilon}(x,y)dydx\\
&= -2\int_{\Omega_1} g(x-y)(u_\delta(x) - u_\delta(y))dydx\\
&\quad - 2\int_{\Omega_2} g(x-y)(u_\delta(x) - u_\delta(y))dydx\\
&=: I_{1,\varepsilon} + I_{2,\varepsilon}\,.
\end{aligned}
$$

Arguing as in case 1, we rewrite $I_{1,\varepsilon}$ as

$$-2\,\text{sgn}(r_{i-1}\varepsilon)\int_{t_{i-1}^{(\delta)}\wedge t_{i-1}^{(\delta+\varepsilon)}}^{t_{i-1}^{(\delta)}\vee t_{i-1}^{(\delta+\varepsilon)}}\int_{-1}^{1}g(x-y)(u_\delta(x)-u_\delta(y))dydx+O(\varepsilon^2)\,.$$

If $r_i\varepsilon<0$, the second integral satisfies

$$\begin{aligned}
I_{2,\varepsilon}=&-2\int_{t_i^{(\delta+\varepsilon)}}^{t_i^{(\delta)}}\int_{-1}^{1}g(x-y)(u_\delta(x)-u_\delta(y))dydx\\
&+2\int_{t_i^{(\delta+\varepsilon)}}^{t_i^{(\delta)}}\int_{t_{i-1}^{(\delta)}}^{t_{i-1}^{(\delta+\varepsilon)}}g(x-y)(u_\delta(x)-u_\delta(y))dydx\\
&+2\int_{t_i^{(\delta+\varepsilon)}}^{t_i^{(\delta)}}\int_{t_i^{(\delta+\varepsilon)}}^{t_i^{(\delta)}}g(x-y)(u_\delta(x)-u_\delta(y))dydx\\
=&\,2\,\text{sgn}(r_i\varepsilon)\int_{t_i^{(\delta)}\wedge t_i^{(\delta+\varepsilon)}}^{t_i^{(\delta)}\vee t_i^{(\delta+\varepsilon)}}\int_{-1}^{1}g(x-y)(u_\delta(x)-u_\delta(y))dydx+O(\varepsilon^2)\,.
\end{aligned}$$

If $r_i\varepsilon=0$, then $t_i^{(\delta+\varepsilon)}=t_i^{(\delta)}$ (cf. (5.3)) which yields $\lambda^2(\Omega_2)=0$. Thus, the left-hand side is zero, which obviously is also true for the integral term on the right-hand side, and the equation also holds in this case. Altogether, we obtain

$$\begin{aligned}
\left(\chi_{(t_{i-1}^{(\delta+\varepsilon)},t_i^{(\delta+\varepsilon)})}-\chi_{(t_{i-1}^{(\delta)},t_i^{(\delta)})},u_\delta\right)_g=&\\
=&-2\,\text{sgn}(r_{i-1}\varepsilon)\int_{t_{i-1}^{(\delta)}\wedge t_{i-1}^{(\delta+\varepsilon)}}^{t_{i-1}^{(\delta)}\vee t_{i-1}^{(\delta+\varepsilon)}}\int_{-1}^{1}g(x-y)(u_\delta(x)-u_\delta(y))dydx\\
&+2\,\text{sgn}(r_i\varepsilon)\int_{t_i^{(\delta)}\wedge t_i^{(\delta+\varepsilon)}}^{t_i^{(\delta)}\vee t_i^{(\delta+\varepsilon)}}\int_{-1}^{1}g(x-y)(u_\delta(x)-u_\delta(y))dydx+O(\varepsilon^2)
\end{aligned}$$

Case 3: $r_{i-1}\varepsilon\le0$, $r_i>0$. This case is treated analogously to case 2, and we obtain the same formula.

Case 4: $r_{i-1}\varepsilon\le0$, $r_i\varepsilon\le0$. Then $t_{i-1}^{(\delta+\varepsilon)}\le t_{i-1}^{(\delta)}<t_i^{(\delta+\varepsilon)}\le t_i^{(\delta)}$ for $\varepsilon\in(-\varepsilon_0,\varepsilon_0)$ and thus

$$\chi_{(t_{i-1}^{(\delta+\varepsilon)},t_i^{(\delta+\varepsilon)})}-\chi_{(t_{i-1}^{(\delta)},t_i^{(\delta)})}=\chi_{(t_{i-1}^{(\delta+\varepsilon)},t_{i-1}^{(\delta)})}-\chi_{(t_i^{(\delta+\varepsilon)},t_i^{(\delta)})}$$

a.e. in $(-1,1)$, i.e.

$$\left(\chi_{(t_{i-1}^{(\delta+\varepsilon)},t_i^{(\delta+\varepsilon)})}-\chi_{(t_{i-1}^{(\delta)},t_i^{(\delta)})},u_\delta\right)_g=\int_{-1}^{1}\int_{-1}^{1}g(x-y)(u_\delta(x)-u_\delta(y))\phi_{\delta,\varepsilon}(x,y)dydx$$

with

$$\phi_{\delta,\varepsilon}(x,y)=\chi_{(t_{i-1}^{(\delta+\varepsilon)},t_{i-1}^{(\delta)})}(x)-\chi_{(t_i^{(\delta+\varepsilon)},t_i^{(\delta)})}(x)-\chi_{(t_{i-1}^{(\delta+\varepsilon)},t_{i-1}^{(\delta)})}(y)+\chi_{(t_i^{(\delta+\varepsilon)},t_i^{(\delta)})}(y)\,.$$

Analogously to case 1, we only have to consider the integral over $\Omega_1 \cup \Omega_2 \cup \Omega_3 \cup \Omega_4 \cup \Omega_5 \cup \Omega_6$, where

$$
\begin{aligned}
\Omega_1 &= (t_{i-1}^{(\delta+\varepsilon)}, t_{i-1}^{(\delta)}) \times (t_i^{(\delta+\varepsilon)}, t_i^{(\delta)}) \\
\Omega_2 &= (t_i^{(\delta+\varepsilon)}, t_i^{(\delta)}) \times (t_{i-1}^{(\delta+\varepsilon)}, t_{i-1}^{(\delta)}) \\
\Omega_3 &= (t_{i-1}^{(\delta+\varepsilon)}, t_{i-1}^{(\delta)}) \times ((-1, t_{i-1}^{(\delta+\varepsilon)}) \cup (t_{i-1}^{(\delta)}, t_i^{(\delta+\varepsilon)}) \cup (t_i^{(\delta)}, 1)) \\
\Omega_4 &= ((-1, t_{i-1}^{(\delta+\varepsilon)}) \cup (t_{i-1}^{(\delta)}, t_i^{(\delta+\varepsilon)}) \cup (t_i^{(\delta)}, 1)) \times (t_{i-1}^{(\delta+\varepsilon)}, t_{i-1}^{(\delta)}) \\
\Omega_5 &= (t_i^{(\delta+\varepsilon)}, t_i^{(\delta)}) \times ((-1, t_{i-1}^{(\delta+\varepsilon)}) \cup (t_{i-1}^{(\delta)}, t_i^{(\delta+\varepsilon)}) \cup (t_i^{(\delta)}, 1)) \\
\Omega_6 &= ((-1, t_{i-1}^{(\delta+\varepsilon)}) \cup (t_{i-1}^{(\delta)}, t_i^{(\delta+\varepsilon)}) \cup (t_i^{(\delta)}, 1)) \times (t_i^{(\delta+\varepsilon)}, t_i^{(\delta)}) \,,
\end{aligned}
$$

since $\phi_{\delta,\varepsilon}$ vanishes almost everywhere else on $[-1,1] \times [-1,1]$. As before, due to the symmetry of the integrand, we may restrict to the domains Ω_1, Ω_3 and Ω_5, taking each integral twice, and we obtain

$$
\begin{aligned}
\left(\chi_{(t_{i-1}^{(\delta+\varepsilon)}, t_i^{(\delta+\varepsilon)})} - \chi_{(t_{i-1}^{(\delta)}, t_i^{(\delta)})}, u_\delta \right)_g &= 4 \int_{\Omega_1} g(x-y)(u_\delta(x) - u_\delta(y)) dy dx \\
&\quad 2 \int_{\Omega_3} g(x-y)(u_\delta(x) - u_\delta(y)) dy dx \\
&\quad - 2 \int_{\Omega_5} g(x-y)(u_\delta(x) - u_\delta(y)) dy dx \\
&=: I_{1,\varepsilon} + I_{2,\varepsilon} + I_{3,\varepsilon} \,,
\end{aligned}
$$

where

$$
|I_{1,\varepsilon}| = 4 \left| \int_{t_{i-1}^{(\delta+\varepsilon)}}^{t_{i-1}^{(\delta)}} \int_{t_i^{(\delta+\varepsilon)}}^{t_i^{(\delta)}} g(x-y)(u_\delta(x) - u_\delta(y)) dy dx \right| \le C\varepsilon^2 \,.
$$

If $r_{i-1}\varepsilon < 0$,

$$
\begin{aligned}
I_{2,\varepsilon} &= 2 \int_{t_{i-1}^{(\delta+\varepsilon)}}^{t_{i-1}^{(\delta)}} \int_{-1}^{1} g(x-y)(u_\delta(x) - u_\delta(y)) dy dx + O(\varepsilon^2) \\
&= -2 \operatorname{sgn}(r_{i-1}\varepsilon) \int_{t_{i-1}^{(\delta)} \wedge t_{i-1}^{(\delta+\varepsilon)}}^{t_{i-1}^{(\delta)} \vee t_{i-1}^{(\delta+\varepsilon)}} \int_{-1}^{1} g(x-y)(u_\delta(x) - u_\delta(y)) dy dx + O(\varepsilon^2) \,,
\end{aligned}
$$

which also holds if $r_{i-1}\varepsilon = 0$ since in that case, $t_{i-1}^{(\delta)} = t_{i-1}^{(\delta+\varepsilon)}$ which implies that both $I_{2,\varepsilon}$ and the integral on the right-hand side are zero. Similarly, if $r_i\varepsilon < 0$,

$$
\begin{aligned}
I_{3,\varepsilon} &= -2 \int_{t_i^{(\delta+\varepsilon)}}^{t_i^{(\delta)}} \int_{-1}^{1} g(x-y)(u_\delta(x) - u_\delta(y)) dy dx + O(\varepsilon^2) \\
&= 2 \operatorname{sgn}(r_i\varepsilon) \int_{t_i^{(\delta)} \wedge t_i^{(\delta+\varepsilon)}}^{t_i^{(\delta)} \vee t_i^{(\delta+\varepsilon)}} \int_{-1}^{1} g(x-y)(u_\delta(x) - u_\delta(y)) dy dx + O(\varepsilon^2)
\end{aligned}
$$

which again is easily extended to the whole case $r_i\varepsilon \le 0$, and we get the same formula as in the previous cases.

Thus we have shown that

$$
\left(\chi_{(t_{i-1}^{(\delta+\varepsilon)}, t_i^{(\delta+\varepsilon)})} - \chi_{(t_{i-1}^{(\delta)}, t_i^{(\delta)})}, u_\delta \right)_g =
$$

$$= -2\operatorname{sgn}(r_{i-1}\varepsilon) \int_{t_{i-1}^{(\delta)} \wedge t_{i-1}^{(\delta+\varepsilon)}}^{t_{i-1}^{(\delta)} \vee t_{i-1}^{(\delta+\varepsilon)}} \int_{-1}^{1} g(x-y)(u_\delta(x) - u_\delta(y))dydx$$

$$+ 2\operatorname{sgn}(r_i\varepsilon) \int_{t_i^{(\delta)} \wedge t_i^{(\delta+\varepsilon)}}^{t_i^{(\delta)} \vee t_i^{(\delta+\varepsilon)}} \int_{-1}^{1} g(x-y)(u_\delta(x) - u_\delta(y))dydx + O(\varepsilon^2)$$

$$=: T_1^{(\varepsilon)} + T_2^{(\varepsilon)} + O(\varepsilon^2). \tag{5.17}$$

In the following, let $\varepsilon \in (-\varepsilon_0, \varepsilon_0) \setminus \{0\}$. If $r_{i-1} \neq 0$, we obtain

$$\frac{1}{\varepsilon}T_1^{(\varepsilon)} = -2r_{i-1}\frac{1}{|r_{i-1}\varepsilon|} \int_{t_{i-1}^{(\delta)} \wedge t_{i-1}^{(\delta+\varepsilon)}}^{t_{i-1}^{(\delta)} \vee t_{i-1}^{(\delta+\varepsilon)}} \int_{-1}^{1} g(x-y)(u_\delta(x) - u_\delta(y))dydx,$$

and since $(t_{i-1}^{(\delta)} \vee t_{i-1}^{(\delta+\varepsilon)}) - (t_{i-1}^{(\delta)} \wedge t_{i-1}^{(\delta+\varepsilon)}) = |r_{i-1}\varepsilon|$ and both $t_{i-1}^{(\delta)} \vee t_{i-1}^{(\delta+\varepsilon)}$ and $t_{i-1}^{(\delta)} \wedge t_{i-1}^{(\delta+\varepsilon)}$ converge to $t_{i-1}^{(\delta)}$ for $\varepsilon \to 0$ (cf. (5.3)), we deduce

$$\lim_{\varepsilon \to 0} \frac{1}{\varepsilon}T_1^{(\varepsilon)} = -2r_{i-1} \int_{-1}^{1} g(t_{i-1}^{(\delta)} - y)(u_\delta(t_{i-1}^{(\delta)}) - u_\delta(y))dy,$$

which is also true if $r_{i-1} = 0$, since in this case, $t_{i-1}^{(\delta)} \vee t_{i-1}^{(\delta+\varepsilon)} = t_{i-1}^{(\delta)} \wedge t_{i-1}^{(\delta+\varepsilon)}$ and thus $T_1^{(\varepsilon)} = 0$, so that both sides of the equation are zero. Analogously, we show that

$$\lim_{\varepsilon \to 0} \frac{1}{\varepsilon}T_2^{(\varepsilon)} = 2r_i \int_{-1}^{1} g(t_i^{(\delta)} - y)(u_\delta(t_i^{(\delta)}) - u_\delta(y))dydx.$$

Setting these into (5.17), we get

$$\lim_{\varepsilon \to 0} \frac{1}{\varepsilon} \left(\chi_{(t_{i-1}^{(\delta+\varepsilon)}, t_i^{(\delta+\varepsilon)})} - \chi_{(t_{i-1}^{(\delta)}, t_i^{(\delta)})}, u_\delta \right)_g = 2r_i \int_{-1}^{1} g(t_i^{(\delta)} - y)(u_\delta(t_i^{(\delta)}) - u_\delta(y))dydx$$

$$- 2r_{i-1} \int_{-1}^{1} g(t_{i-1}^{(\delta)} - y)(u_\delta(t_{i-1}^{(\delta)}) - u_\delta(y))dydx,$$

which together with (5.15) and (5.16) yields

$$(h_{u,r}^{(i)})'(\delta) = \lim_{\varepsilon \to 0} \frac{1}{\varepsilon}(h_{u,r}^{(i)}(\delta + \varepsilon) - h_{u,r}^{(i)}(\delta)) = 2\sum_{k=1}^{N} \alpha_k \left(\chi_{(t_{k-1}^{(\delta)}, t_k^{(\delta)})}, \chi_{(t_{i-1}^{(\delta)}, t_i^{(\delta)})} \right)_g$$

$$+ 2r_i \int_{-1}^{1} g(t_i^{(\delta)} - y)(u_\delta(t_i^{(\delta)}) - u_\delta(y))dy \tag{5.18}$$

$$- 2r_{i-1} \int_{-1}^{1} g(t_{i-1}^{(\delta)} - y)(u_\delta(t_{i-1}^{(\delta)}) - u_\delta(y))dy$$

which shows the differentiability and (5.14). To prove the continuity of each $(h_{u,r}^{(i)})'$ in every $\delta \in (-\delta_0, \delta_0)$, we show that

$$\lim_{\varepsilon \to 0} u_{\delta+\varepsilon}(t_i^{(\delta+\varepsilon)}) = u_\delta(t_i^{(\delta)})$$

for every $\delta \in (-\delta_0, \delta_0)$ and every $i = 0, \ldots, N$. This is obvious for $i = 0$, since $t_0^{(\delta+\varepsilon)} = t_0^{(\delta)} = -1$ and $u_{\delta+\varepsilon}(-1) = u_\delta(-1)$. From (5.5) and (5.3), for $i = 1, \ldots, N$ we obtain

$$
\begin{aligned}
u_{\delta+\varepsilon}(t_i^{(\delta+\varepsilon)}) - u_\delta(t_i^{(\delta)}) &= \int_{-1}^{t_i^{(\delta+\varepsilon)}} u'_{\delta+\varepsilon}(x)dx - \int_{-1}^{t_i^{(\delta)}} u'_\delta(x)dx \\
&= \sum_{j=1}^{i} \left[\int_{t_{j-1}^{(\delta+\varepsilon)}}^{t_j^{(\delta+\varepsilon)}} (-1)^{j-1}dx - \int_{t_{j-1}^{(\delta)}}^{t_j^{(\delta)}} (-1)^{j-1}dx \right] \\
&= \sum_{j=1}^{i} (-1)^{j-1}(t_j^{(\delta+\varepsilon)} - t_{j-1}^{(\delta+\varepsilon)} - t_j^{(\delta)} + t_{j-1}^{(\delta)}) \\
&= \sum_{j=1}^{i} (-1)^{j-1}(\varepsilon r_j - \varepsilon r_{j-1}),
\end{aligned}
$$

hence

$$
|u_{\delta+\varepsilon}(t_i^{(\delta+\varepsilon)}) - u_\delta(t_i^{(\delta)})| \le |\varepsilon| \sum_{j=1}^{i} (-1)^{j-1}|r_j - r_{j-1}| \to 0 \quad \text{for} \quad \varepsilon \to 0.
$$

The rest now easily follows from Lebesgue's convergence theorem: g is bounded and $g(t_i^{(\delta+\varepsilon)} - y) \to g(t_i^{(\delta)} - y)$ for $\varepsilon \to 0$ due to the continuity of g and since $t_i^{(\delta+\varepsilon)} \to t_i^{(\delta)}$ for $\varepsilon \to 0$. Furthermore, $u_{\delta+\varepsilon} \to u_\delta$ pointwise a.e., and u_δ are (essentially) uniformly bounded (cf. Lemma 5.1). Obviously,

$$
\chi_{(t_{k-1}^{(\delta+\varepsilon)}, t_k^{(\delta+\varepsilon)})} \to \chi_{(t_{k-1}^{(\delta)}, t_k^{(\delta)})}
$$

pointwise a.e. for $\varepsilon \to 0$, which implies that the first line in (5.18) is also continuous in δ.

$\square$

Chapter 6

The Equidistant Case

6.1 Sawtooth Functions with Equidistant Corners

In order to identify certain local minimizers of E in the set of all $u \in \mathcal{S}^0_{per}(-1,1)$ with $\#(Su' \cap [-1,1)) = N$, where N is a fixed even number, we apply the results of the previous chapter to these particular sawtooth functions. The functions that we conjecture to be local minimizers are characterized by their corners lying equidistantly in $[-1,1)$. Thus, for a fixed even N, denoting the number of corners in $[-1,1)$, we define

$$\bar{t}_i = -1 + \frac{2}{N}i \quad \text{for} \quad i = 0, \ldots, N \tag{6.1}$$

as equidistant points in $[-1,1]$. In particular we have

$$-1 = \bar{t}_0 < \bar{t}_1 < \cdots < \bar{t}_{N-1} < \bar{t}_N = 1 \, .$$

We define $v_N \in BV((-1,1), \{-1,1\})$ by

$$v_N(x) = (-1)^{i-1} \quad \text{for} \quad x \in (\bar{t}_{i-1}, \bar{t}_i), \quad i = 1, \ldots, N \tag{6.2}$$

and $\bar{u}_N \in \mathcal{S}(-1,1)$ by

$$\bar{u}_N(x) = \int_{-1}^{x} v_N(\xi)d\xi - \frac{1}{N}. \tag{6.3}$$

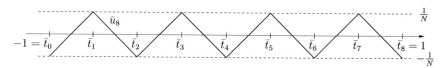

Figure 6.1: $\bar{u}_N$ for $N = 8$

Remark: We omit the dependence of the corners $\bar{t}_i$ on N, since it would lead to confusing notation.

$\bar{u}_N$ has the following properties:

Lemma 6.1 $\bar{u}_N \in \mathcal{S}^0_{per}(-1,1)$ *with*

(i) $\bar{u}'_N(x) = (-1)^{i-1}$ *for* $x \in (\bar{t}_{i-1}, \bar{t}_i)$, $i = 1, \ldots, N$.

(ii) $\bar{u}_N(-1) = \bar{u}_N(1) = -\frac{1}{N}$, *and* $\bar{u}_N(\bar{t}_k) = (-1)^{k-1}\frac{1}{N}$ *for* $k = 0, \ldots, N$

(iii) $\int_{\bar{t}_{i-1}}^{\bar{t}_i} \bar{u}_N(x)dx = 0$ *for* $i = 1, \ldots, N$

(iv) $\bar{u}_N(x + \frac{2}{N}) = -\bar{u}_N(x)$ *for every* $x \in \mathbb{R}$

(v) $\bar{u}_N(x) = \bar{u}_N(-x)$ *for every* $x \in \mathbb{R}$

(vi) $|\bar{u}_N(x)| \leq \frac{1}{N}$ *for every* $x \in \mathbb{R}$

(vii) $\bar{u}_N(x + \bar{t}_{i-1}) = -\bar{u}_N(\bar{t}_i - x)$ *for every* $x \in \mathbb{R}$

where $\bar{u}_N$ *also denotes the 2-periodic extension of* $\bar{u}_N$ *on* $\mathbb{R}$.

Proof: We first note that $\bar{u}_N \in \mathcal{S}^0_{per}(-1,1)$ follows directly from (ii) and (iii). (i) is immediately deduced from (6.2) and (6.3). To show (ii), we first note that $\bar{t}_i - \bar{t}_{i-1} = \frac{2}{N}$ due to (6.1). Using (6.3) and (6.2) we deduce

$$\bar{u}_N(\bar{t}_k) = \int_{-1}^{\bar{t}_k} v_N(\xi)d\xi - \frac{1}{N} = \sum_{i=1}^{k} \int_{\bar{t}_{i-1}}^{\bar{t}_i} (-1)^{i-1}d\xi - \frac{1}{N} = \frac{2}{N}\sum_{i=1}^{k}(-1)^{i-1} - \frac{1}{N} = (-1)^{k-1}\frac{1}{N}$$

for $k = 1, \ldots, N$, where the last step follows by considering the cases k even / k odd separately. Since N is even, this implies $\bar{u}_N(1) = \bar{u}_N(\bar{t}_N) = -\frac{1}{N}$, while $\bar{u}_N(\bar{t}_0) = \bar{u}_N(-1) = -\frac{1}{N}$ immediately follows from the definition of $\bar{u}_N$. To prove (iii), we again use above definitions and $\bar{t}_i - \bar{t}_{i-1} = \frac{2}{N}$ to obtain

$$\int_{\bar{t}_{i-1}}^{\bar{t}_i} \bar{u}_N(x)dx = \int_{\bar{t}_{i-1}}^{\bar{t}_i} \int_{-1}^{x} v_N(\xi)d\xi dx - \frac{2}{N^2}$$

$$= \sum_{j=1}^{i-1} \int_{\bar{t}_{i-1}}^{\bar{t}_i} \int_{\bar{t}_{j-1}}^{\bar{t}_j} (-1)^{j-1}d\xi dx + \int_{\bar{t}_{i-1}}^{\bar{t}_i} \int_{\bar{t}_{i-1}}^{x} (-1)^{i-1}d\xi dx - \frac{2}{N^2}$$

$$= \frac{4}{N^2}\sum_{j=1}^{i-1}(-1)^{j-1} + \frac{1}{2}(-1)^{i-1}(\bar{t}_i - \bar{t}_{i-1})^2 - \frac{2}{N^2}$$

$$= \frac{4}{N^2}\sum_{j=1}^{i-1}(-1)^{j-1} + (-1)^{i-1}\frac{2}{N^2} - \frac{2}{N^2}.$$

If i is even, we deduce

$$\int_{\bar{t}_{i-1}}^{\bar{t}_i} \bar{u}_N(x)dx = \frac{4}{N^2} - \frac{2}{N^2} - \frac{2}{N^2} = 0.$$

If i is odd, we get

$$\int_{\bar{t}_{i-1}}^{\bar{t}_i} \bar{u}_N(x)dx = 0 + \frac{2}{N^2} - \frac{2}{N^2} = 0,$$

which shows (iii). To prove (iv), we note that the periodic extension of v_N to $\mathbb{R}$ satisfies $v_N(x + \frac{2}{N}) = -v_N(x)$ a.e. on $\mathbb{R}$, which follows from (6.2), since $\bar{t}_i - \bar{t}_{i-1} = \frac{2}{N}$ for $i = 1, \ldots, N$ and N is an even number. Thus

$$\begin{aligned}
\bar{u}_N(x) + \bar{u}_N(x + \frac{2}{N}) &= \int_{-1}^{x} v_N(\xi)d\xi + \int_{-1}^{x+\frac{2}{N}} v_N(\xi)d\xi - \frac{2}{N} \\
&= \int_{-1}^{x} v_N(\xi)d\xi - \int_{-1+\frac{2}{N}}^{x} v_N(\xi)d\xi - \frac{2}{N} \\
&= -\int_{-1-\frac{2}{N}}^{-1} v_N(\xi)d\xi - \frac{2}{N} = -\int_{1-\frac{2}{N}}^{1} v_N(\xi)d\xi - \frac{2}{N} \\
&= -\int_{\bar{t}_{N-1}}^{\bar{t}_N} (-1)^{N-1}d\xi - \frac{2}{N} = \bar{t}_N - \bar{t}_{N-1} - \frac{2}{N} = \frac{2}{N} - \frac{2}{N} = 0
\end{aligned}$$

for $x \in [-1, 1 - \frac{2}{N}]$. Thus we have shown the $\frac{2}{N}$-antiperiodicity of $\bar{u}_N$ on $[-1, 1]$, which is easily extended to $\mathbb{R}$, since $[-1, 1]$ consists of an even number of intervals $[\bar{t}_{i-1}, \bar{t}_i]$ of length $\frac{2}{N}$. As for (v), it is sufficient to show the equation for $x \in (0, 1)$. Since $\bar{t}_{\frac{N}{2}} = 0$, for every such x (up to a finite set) we find $i \in \{\frac{N}{2} + 1, \ldots, N\}$ with $x \in (\bar{t}_{i-1}, \bar{t}_i)$, so that

$$-x \in (-\bar{t}_i, -\bar{t}_{i-1}) = (1 - \frac{2}{N}i, 1 - \frac{2}{N}(i-1)) = (-1 + \frac{2}{N}(N-i), -1 + \frac{2}{N}(N-i+1)),$$

i.e. $-x \in (\bar{t}_{N-i}, \bar{t}_{N-i+1})$, and since N is even we obtain

$$v_N(-x) = (-1)^{N-i} = (-1)^i = -v_N(x).$$

Thus, v_N is, up to a finite set, anti-symmetric on $[-1, 1]$, and we get

$$\bar{u}_N(x) - \bar{u}_N(-x) = \int_{-1}^{x} v_N(\xi)d\xi - \int_{-1}^{-x} v_N(\xi)d\xi = \int_{-x}^{x} v_N(\xi)d\xi = 0,$$

and this identity is immediately transferred to the periodic extension $\bar{u}_N$. To show (vi), for $x \in [\bar{t}_{i-1}, \bar{t}_i]$, $i \in \{1, \ldots, N\}$ we get

$$\begin{aligned}
\bar{u}_N(x) &= \sum_{j=1}^{i-1} \int_{\bar{t}_{j-1}}^{\bar{t}_j} (-1)^{j-1}d\xi + \int_{\bar{t}_{i-1}}^{x} (-1)^{i-1}d\xi - \frac{1}{N} \\
&= \frac{2}{N} \sum_{j=1}^{i-1}(-1)^{j-1} + (-1)^{i-1}(x - \bar{t}_{i-1}) - \frac{1}{N}.
\end{aligned}$$

Thus, if i is even, we deduce

$$\bar{u}_N(x) = \frac{1}{N} - x + \bar{t}_{i-1} \in [\frac{1}{N} - \bar{t}_i + \bar{t}_{i-1}, \frac{1}{N}] = [-\frac{1}{N}, \frac{1}{N}]$$

and if i is odd, we conclude

$$\bar{u}_N(x) = x - \bar{t}_{i-1} - \frac{1}{N} \in [-\frac{1}{N}, \bar{t}_i - \bar{t}_{i-1} - \frac{1}{N}] = [-\frac{1}{N}, \frac{1}{N}],$$

so that $|\bar{u}_N| \le \frac{1}{N}$ on $[-1, 1]$, which extends to the whole real line. To prove (vii), we use (v) and (iv) as well as the 2-periodicity to obtain

$$\begin{aligned}
\bar{u}_N(x + \bar{t}_{i-1}) &= \bar{u}_N(-x - \bar{t}_{i-1}) = \bar{u}_N(-x + 1 - \frac{2}{N}(i-1)) \\
&= \bar{u}_N(-x - 1 - \frac{2}{N}(i-1)) = -\bar{u}_N(-x - 1 - \frac{2}{N}(i-1) + \frac{2}{N}(2i-1)) \\
&= -\bar{u}_N(-x - 1 + \frac{2}{N}i) = -\bar{u}_N(\bar{t}_i - x)
\end{aligned}$$

for $x \in \mathbb{R}$, and the proof is complete. $\qquad\qquad\qquad\qquad\qquad\qquad\qquad\qquad\square$

6.2   Criticality

As in the previous chapter, we fix $(r_1, \ldots, r_{N-1}) \in \mathbb{R}^{N-1}$ and define the corresponding perturbations $\bar{u}_{N,\delta}$. According to the definitions at the beginning of chapter 5, for $\delta \in (-\delta_0, \delta_0)$ we set

$$\bar{t}_i^{(\delta)} = \bar{t}_i + r_i \delta \quad \text{for} \quad i = 0, \ldots, N, \tag{6.4}$$

where $r_0 = r_n = 0$ is set, i.e. $\bar{t}_0^{(\delta)} = -1$, $\bar{t}_N^{(\delta)} = 1$. As before, $\delta_0 > 0$ is chosen such that the monotonicity

$$-1 = \bar{t}_0^{(\delta)} < \bar{t}_1^{(\delta)} < \cdots < \bar{t}_{N-1}^{(\delta)} < \bar{t}_N^{(\delta)} = 1$$

holds for every $\delta \in (-\delta_0, \delta_0)$. The "$r$-perturbation" of $\bar{u}_N$ according to (5.5) is obviously given by

$$\bar{u}_{N,\delta}(x) = \int_{-1}^{x} v_{N,\delta}(\xi)d\xi - \frac{1}{N} \tag{6.5}$$

for $x \in [-1, 1]$, where

$$v_{N,\delta}(x) = (-1)^{i-1} \quad \text{for} \quad x \in (\bar{t}_{i-1}^{(\delta)}, \bar{t}_i^{(\delta)}), \quad i = 1, \ldots, N. \tag{6.6}$$

In particular, we have

$$\bar{u}_{N,\delta}'(x) = (-1)^{i-1} \quad \text{for} \quad x \in (\bar{t}_{i-1}^{(\delta)}, \bar{t}_i^{(\delta)}), \quad i = 1, \ldots, N. \tag{6.7}$$

As before, setting $\delta = 0$ yields $\bar{u}_{N,0} = \bar{u}_N$. Although it is clear that $\bar{u}_{N,\delta}$ is a sawtooth function with slope ± 1 and $N - 1$ inner corners, $\bar{u}_{N,\delta}$ might not be periodic if $\delta \ne 0$.

Lemma 6.2 *Let $\delta \in (-\delta_0, \delta_0) \setminus \{0\}$, where $\delta_0 > 0$ is chosen as above. Then $\bar{u}_{N,\delta}$ as given by (6.5) and (6.6) is periodic (i.e. $\bar{u}_{N,\delta}(-1) = \bar{u}_{N,\delta}(-1)$) if and only if*

$$\sum_{i=1}^{N-1} (-1)^{i-1} r_i = 0. \tag{6.8}$$

Proof: Periodicity holds if and only if $\bar{u}_{N,\delta}(1) = \bar{u}_{N,\delta}(-1) = -\frac{1}{N}$, i.e.

$$0 = \int_{-1}^{1} \bar{u}'_{N,\delta}(\xi)d\xi = \sum_{i=1}^{N} \int_{\bar{t}_{i-1}^{(\delta)}}^{\bar{t}_{i}^{(\delta)}} (-1)^{i-1} d\xi = \sum_{i=1}^{N} (-1)^{i-1} \left(\bar{t}_i - \bar{t}_{i-1} + (r_i - r_{i-1})\delta \right)$$

$$= \sum_{i=1}^{N} (-1)^{i-1} \left(\frac{2}{N} + (r_i - r_{i-1})\delta \right) = \delta \sum_{i=1}^{N} (-1)^{i-1}(r_i - r_{i-1}) = 2\delta \sum_{i=1}^{N-1} (-1)^{i-1} r_i \,,$$

where we used the fact that N is even and $r_0 = r_N = 0$. Since $\delta \neq 0$, the condition now follows.   □

We recall the definition (5.7) of the bilinear form associated with g. According to (5.6), we define $\bar{F}_r : (-\delta_0, \delta_0) \to \mathbb{R}$ by

$$\bar{F}_r(\delta) = F_{\bar{u}_{N},r}(\delta) = E(\bar{u}_{N,\delta})\,. \tag{6.9}$$

As we want to find out under which conditions $\bar{u}_N$ is a local minimizer of E in the set $\mathcal{S}^0_{per}(-1,1)$, we will make use of the results in the previous chapter in order to examine the derivatives of the function $\bar{F}_r$ which describes the change of the nonlocal energy E under the respective perturbation of $\bar{u}_N$. As for the first derivative, the following holds:

Theorem 6.3 *Let $(r_1, \ldots, r_{N-1}) \in \mathbb{R}^{N-1}$, and choose $\delta_0 > 0$ as above. For every $\delta \in (-\delta_0, \delta_0)$, let $\bar{u}_{N,\delta}$ as defined by (6.5) and (6.6). Then $\bar{F}_r : (-\delta_0, \delta_0) \to \mathbb{R}$ as given by (6.9) is differentiable, and*

$$\bar{F}'_r(0) = 0. \tag{6.10}$$

Proof: Due to (6.9), the differentiability on $(-\delta_0, \delta_0)$ follows directly from Theorem 5.2, which also yields

$$\bar{F}'_r(0) = F'_{\bar{u}_N,r}(0) = 4 \sum_{i=1}^{N} \alpha_i (\chi_{(\bar{t}_{i-1},\bar{t}_i)}, \bar{u}_N)_g,$$

with $\alpha_1, \ldots, \alpha_N$ as defined in (5.11). We aim to show that

$$(\chi_{(\bar{t}_{i-1},\bar{t}_i)}, \bar{u}_N)_g = 0$$

for $i = 1, \ldots, N$, which immediately yields (6.10). From the symmetry of g, we derive that the function $\varphi(x,y) = g(x-y)(\bar{u}_N(x) - \bar{u}_N(y))$ satisfies $\varphi(y,x) = -\varphi(x,y)$, which implies

$$\int_{\bar{t}_{i-1}}^{\bar{t}_i} \int_{\bar{t}_{i-1}}^{\bar{t}_i} g(x-y)(\bar{u}_N(x) - \bar{u}_N(y))dxdy = 0.$$

Thus, using the symmetry and the 2-periodicity of g,

$$(\chi_{(\bar{t}_{i-1},\bar{t}_i)}, \bar{u}_N)_g = \int_{-1}^{1} \int_{-1}^{1} g(x-y)(\chi_{(\bar{t}_{i-1},\bar{t}_i)}(x) - \chi_{(\bar{t}_{i-1},\bar{t}_i)}(y))(\bar{u}_N(x) - \bar{u}_N(y))dxdy$$

$$=2 \int_{-1}^{\bar{t}_{i-1}} \int_{\bar{t}_{i-1}}^{\bar{t}_i} g(x-y)(\bar{u}_N(x) - \bar{u}_N(y))dxdy$$

$$+ 2 \int_{\bar{t}_i}^1 \int_{\bar{t}_{i-1}}^{\bar{t}_i} g(x-y)(\bar{u}_N(x) - \bar{u}_N(y))dxdy$$

$$=2 \int_{-1}^1 \int_{\bar{t}_{i-1}}^{\bar{t}_i} g(x-y)(\bar{u}_N(x) - \bar{u}_N(y))dxdy$$

$$=2 \int_{\bar{t}_{i-1}}^{\bar{t}_i} \bar{u}_N(x) \int_{-1}^1 g(x-y)dydx - 2 \int_{\bar{t}_{i-1}}^{\bar{t}_i} \int_{-1}^1 g(x-y)\bar{u}_N(y)dydx$$

$$=2 \int_{\bar{t}_{i-1}}^{\bar{t}_i} \bar{u}_N(x)dx \int_{-1}^1 g(y)dy - 2 \int_{\bar{t}_{i-1}}^{\bar{t}_i} \int_{-1}^1 g(y-x)\bar{u}_N(y)dydx$$

$$=-2 \int_{\bar{t}_{i-1}}^{\bar{t}_i} \int_{-1}^1 g(y-x)\bar{u}_N(y)dydx,$$

since $\int_{\bar{t}_{i-1}}^{\bar{t}_i} \bar{u}_N(x)dx = 0$ due to Lemma 6.1. We claim that

$$x \mapsto \int_{-1}^1 g(y-x)\bar{u}_N(y)dy$$

is skew symmetric on $(\bar{t}_{i-1}, \bar{t}_i)$, which would imply $(\chi_{(\bar{t}_{i-1},\bar{t}_i)}, \bar{u}_N)_g = 0$. To show this, we set $x_1 = \bar{t}_{i-1} + \beta$, $x_2 = \bar{t}_i - \beta$ with $\beta \in [0, \frac{\bar{t}_{i-1}+\bar{t}_i}{2}]$. Then, using the 2-periodicity of $\bar{u}_N$ and g, Lemma 6.1 (vii) and (v) and the symmetry of g, we obtain

$$\int_{-1}^1 g(y-x_1)\bar{u}_N(y)dy = \int_{-1}^1 g(y)\bar{u}_N(y+x_1)dy = \int_{-1}^1 g(y)\bar{u}_N(y+\bar{t}_{i-1}+\beta)dy$$

$$= -\int_{-1}^1 g(y)\bar{u}_N(\bar{t}_i - \beta - y)dy = -\int_{-1}^1 g(y)\bar{u}_N(x_2-y)dy$$

$$= -\int_{-1}^1 g(y)\bar{u}_N(x_2+y)dy = -\int_{-1}^1 g(y-x_2)\bar{u}_N(y)dy,$$

which implies the skew symmetry, so that the proof is complete. $\qquad \square$

6.3 The "Hessian" Matrix A_N

Since we are interested in local minimizers under periodic boundary conditions, we will consider the case where $\bar{u}_{N,\delta}$ is periodic, which due to (6.8) and (5.11) is equivalent to $\alpha_N = 0$. Furthermore, we always have $\alpha_1 = 0$ (see (5.11)). By applying Theorem 5.3, the second derivative of $\bar{F}_r$ is of a special form, described in the next Theorem.

Theorem 6.4 *For $N \in \mathbb{N}$ even, let $\bar{u}_N \in \mathcal{S}_{per}^0(-1,1)$ as defined by (6.3), (6.2). Let $(r_1, \ldots, r_{N-1}) \in \mathbb{R}^{N-1}$ such that the periodicity condition (6.8) holds, and choose $\delta_0 > 0$ as before. Let $\bar{F}_r : (-\delta_0, \delta_0) \to \mathbb{R}$ as given in (6.9). Let $\alpha = (\alpha_1, \ldots, \alpha_N)^t \in \mathbb{R}^N$ as defined in (5.11) (in particular, $\alpha_1 = \alpha_N = 0$). Then $\bar{F}_r \in C^2(-\delta_0, \delta_0)$ with*

$$\frac{1}{8}\bar{F}_r''(0) = \frac{2}{N}\alpha^t A_N \alpha \, ,$$

where

$$A_N = B_N + C_N + D_N \in \mathbb{R}^{N \times N}$$

with $B_N = (b_{|i-j|}^{(N)})_{i,j=1,\ldots,N}$, $C_N = (c_{|i-j|}^{(N)})_{i,j=1,\ldots,N}$, $D_N = (d_{|i-j|}^{(N)})_{i,j=1,\ldots,N} \in \mathbb{R}^{N \times N}$ being symmetric and circulant, i.e.

$$b_k^{(N)} = b_{N-k}^{(N)}, \quad c_k^{(N)} = c_{N-k}^{(N)}, \quad d_k^{(N)} = d_{N-k}^{(N)}$$

for $k = 1, \ldots, N-1$, the coefficients of B_N and D_N given by

$$b_0^{(N)} = \int_{-1}^{1} g(y)dy, \quad b_1^{(N)} = b_{N-1}^{(N)} = \frac{1}{2}b_0^{(N)}, \quad b_2^{(N)} = \cdots = b_{N-2}^{(N)} = 0,$$

$$d_0^{(N)} = N(-1)^{\frac{N}{2}+1}\int_{-1}^{1} g(y)\bar{u}_N(y)dy, \quad d_1^{(N)} = d_{N-1}^{(N)} = -\frac{1}{2}d_0^{(N)}, \tag{6.11}$$

$$d_2^{(N)} = \cdots = d_{N-2}^{(N)} = 0,$$

while those of C_N are

$$c_k^{(N)} = -N\int_0^{\frac{2}{N}}\int_{\frac{2}{N}k}^{\frac{2}{N}(k+1)} g(x-y)dydx \tag{6.12}$$

for $k = 0, \ldots, N-1$. Alternatively, $C_N = (c_{ij}^{(N)})_{i,j=1,\ldots,N} \in \mathbb{R}^{N \times N}$ with

$$c_{ij}^{(N)} = -N\int_{\frac{2}{N}(i-1)}^{\frac{2}{N}i}\int_{\frac{2}{N}(j-1)}^{\frac{2}{N}j} g(x-y)dydx \tag{6.13}$$

for $i, j = 1, \ldots, N$.

In particular, $A_N = (a_{|i-j|}^{(N)})_{i,j=1,\ldots,N}$ is symmetric and circulant, i.e. $a_k^{(N)} = a_{N-k}^{(N)}$ for $k = 1, \ldots, N-1$, and its coefficients are

$$\begin{aligned}
a_0^{(N)} &= b_0^{(N)} + c_0^{(N)} + d_0^{(N)} \\
a_1^{(N)} = a_{N-1}^{(N)} &= b_1^{(N)} + c_1^{(N)} + d_1^{(N)} = \frac{1}{2}(b_0^{(N)} - d_0^{(N)}) + c_1^{(N)} \\
a_k^{(N)} &= c_k^{(N)} \quad for \quad k = 2, \ldots, N-2 \, .
\end{aligned} \tag{6.14}$$

Proof: Since $\bar{F}_r = F_{\bar{u}_N,r}$, the regularity follows directly from Theorem 5.3. Obviously, $\alpha_1 = \alpha_N = 0$ (cf.(5.11), (6.8)), and due to (5.13), we obtain, recalling (6.4) and $\bar{u}_{N,0} = \bar{u}_N$,

$$\frac{1}{8}\bar{F}_r''(0) = \frac{1}{8}F_{\bar{u}_N,r}''(0) = \sum_{i=2}^{N-1}\sum_{k=2}^{N-1}\alpha_i\alpha_k\left(\chi_{(\bar{t}_{k-1},\bar{t}_k)},\chi_{(\bar{t}_{i-1},\bar{t}_i)}\right)_g$$

$$+ \sum_{i=2}^{N-1} \alpha_i r_i \int_{-1}^{1} g(\bar{t}_i - y)(\bar{u}_N(\bar{t}_i) - \bar{u}_N(y))dy \tag{6.15}$$

$$- \sum_{i=2}^{N-1} \alpha_i r_{i-1} \int_{-1}^{1} g(\bar{t}_{i-1} - y)(\bar{u}_N(\bar{t}_{i-1}) - \bar{u}_N(y))dy$$

We consider the first term: Let $i = k$, then

$$\left(\chi_{(\bar{t}_{i-1},\bar{t}_i)}, \chi_{(\bar{t}_{i-1},\bar{t}_i)}\right)_g = \int_{-1}^{1} \int_{-1}^{1} g(x-y)(\chi_{(\bar{t}_{i-1},\bar{t}_i)}(x) - \chi_{(\bar{t}_{i-1},\bar{t}_i)}(y))^2 dy dx \,,$$

where we will only have to integrate over $(x,y) \in (\bar{t}_{i-1}, \bar{t}_i) \times ((-1,1) \setminus (\bar{t}_{i-1}, \bar{t}_i))$ and $(x,y) \in ((-1,1) \setminus (\bar{t}_{i-1}, \bar{t}_i)) \times (\bar{t}_{i-1}, \bar{t}_i)$. These two sets are disjoint, and due to the symmetry of the integrand, we would receive the same results for either set, so that we may restrict to the first one and thus obtain

$$
\begin{aligned}
\left(\chi_{(\bar{t}_{i-1},\bar{t}_i)}, \chi_{(\bar{t}_{i-1},\bar{t}_i)}\right)_g &= 2 \int_{\bar{t}_{i-1}}^{\bar{t}_i} \int_{-1}^{\bar{t}_{i-1}} g(x-y)dy dx + 2 \int_{\bar{t}_{i-1}}^{\bar{t}_i} \int_{\bar{t}_i}^{1} g(x-y)dy dx \\
&= 2 \int_{\bar{t}_{i-1}}^{\bar{t}_i} \int_{-1}^{1} g(x-y)dy dx - 2 \int_{\bar{t}_{i-1}}^{\bar{t}_i} \int_{\bar{t}_{i-1}}^{\bar{t}_i} g(x-y)dy dx \\
&= 2 \int_{\bar{t}_{i-1}}^{\bar{t}_i} \int_{-1}^{1} g(y)dy dx - 2 \int_{\bar{t}_{i-1}}^{\bar{t}_i} \int_{\bar{t}_{i-1}}^{\bar{t}_i} g(x-y)dy dx \\
&= \frac{4}{N} \int_{-1}^{1} g(y)dy - 2 \int_{\bar{t}_{i-1}}^{\bar{t}_i} \int_{\bar{t}_{i-1}}^{\bar{t}_i} g(x-y)dy dx,
\end{aligned}
$$

where we made use of the 2-periodicity of g in the third step. For $i \neq k$, we obtain

$$\left(\chi_{(\bar{t}_{k-1},\bar{t}_k)}, \chi_{(\bar{t}_{i-1},\bar{t}_i)}\right)_g = \int_{-1}^{1} \int_{-1}^{1} g(x-y)(\chi_{(\bar{t}_{k-1},\bar{t}_k)}(x) - \chi_{(\bar{t}_{k-1},\bar{t}_k)}(y)) \times$$
$$\times (\chi_{(\bar{t}_{i-1},\bar{t}_i)}(x) - \chi_{(\bar{t}_{i-1},\bar{t}_i)}(y))dy dx \,.$$

Due to $(\bar{t}_{i-1}, \bar{t}_i)$ and $(\bar{t}_{k-1}, \bar{t}_k)$ being disjoint, we easily see that we only have to integrate over $(\bar{t}_{i-1}, \bar{t}_i) \times (\bar{t}_{k-1}, \bar{t}_k)$ and $(\bar{t}_{k-1}, \bar{t}_k) \times (\bar{t}_{i-1}, \bar{t}_i)$. Again, we can restrict to the first one, taking the respective integral twice:

$$\left(\chi_{(\bar{t}_{k-1},\bar{t}_k)}, \chi_{(\bar{t}_{i-1},\bar{t}_i)}\right)_g = -2 \int_{\bar{t}_{i-1}}^{\bar{t}_i} \int_{\bar{t}_{k-1}}^{\bar{t}_k} g(x-y)dy dx \,.$$

Thus, the first line in (6.15) can be written as

$$
\begin{aligned}
\sum_{i=2}^{N-1} \sum_{k=2}^{N-1} \alpha_i \alpha_k \left(\chi_{(\bar{t}_{k-1},\bar{t}_k)}, \chi_{(\bar{t}_{i-1},\bar{t}_i)}\right)_g &= \\
= \frac{4}{N} \sum_{i=2}^{N-1} \alpha_i^2 \int_{-1}^{1} g(y)dy &- 2 \sum_{i=2}^{N-1} \sum_{k=2}^{N-1} \alpha_i \alpha_k \int_{\bar{t}_{i-1}}^{\bar{t}_i} \int_{\bar{t}_{k-1}}^{\bar{t}_k} g(x-y)dy dx \,.
\end{aligned}
\tag{6.16}
$$

As for the remaining terms in (6.15), we use Lemma 6.1 (iv) and the fact that N is even to obtain

$$\bar{u}_N(y + \bar{t}_i) = \bar{u}_N(y - 1 + \frac{2}{N}i) = \bar{u}_N(y + \frac{2}{N}(i - \frac{N}{2})) = (-1)^{i + \frac{N}{2}} \bar{u}_N(y),$$

where $\bar{u}_N$ also denotes the 2-periodic extension of $\bar{u}_N$ to $\mathbb{R}$. Due to the symmetry of g and the 2-periodicity of $\bar{u}_N$ and g, we can write the second and third line in (6.15) as

$$\sum_{i=2}^{N-1} \alpha_i r_i \int_{-1}^{1} g(\bar{t}_i - y)(\bar{u}_N(\bar{t}_i) - \bar{u}_N(y)) dy - \sum_{i=2}^{N-1} \alpha_i r_{i-1} \int_{-1}^{1} g(\bar{t}_{i-1} - y)(\bar{u}_N(\bar{t}_{i-1}) - \bar{u}_N(y)) dy$$

$$= \sum_{i=2}^{N-1} \alpha_i \left(r_{i-1} \int_{-1}^{1} g(y - \bar{t}_{i-1}) \bar{u}_N(y) dy - r_i \int_{-1}^{1} g(y - \bar{t}_i) \bar{u}_N(y) dy \right)$$

$$+ \sum_{i=2}^{N-1} \alpha_i \left(r_i \bar{u}_N(\bar{t}_i) \int_{-1}^{1} g(y - \bar{t}_i) dy - r_{i-1} \bar{u}_N(\bar{t}_{i-1}) \int_{-1}^{1} g(y - \bar{t}_{i-1}) dy \right)$$

$$= \sum_{i=2}^{N-1} \alpha_i \left(r_{i-1} \int_{-1}^{1} g(y) \bar{u}_N(y + \bar{t}_{i-1}) dy - r_i \int_{-1}^{1} g(y) \bar{u}_N(y + \bar{t}_i) dy \right)$$

$$+ \sum_{i=2}^{N-1} \alpha_i \left(r_i \bar{u}_N(\bar{t}_i) - r_{i-1} \bar{u}_N(\bar{t}_{i-1}) \right) \int_{-1}^{1} g(y) dy$$

$$= \sum_{i=2}^{N-1} \alpha_i (-1)^{i + \frac{N}{2} + 1} (r_{i-1} + r_i) \int_{-1}^{1} g(y) \bar{u}_N(y) dy + \frac{1}{N} \sum_{i=2}^{N-1} (-1)^{i-1} \alpha_i (r_{i-1} + r_i) \int_{-1}^{1} g(y) dy,$$

where we have used Lemma 6.1 (ii) in the last step. Using $\alpha_i - \alpha_{i-1} = (-1)^i r_{i-1}$, $\alpha_i - \alpha_{i+1} = (-1)^i r_i$ (which follows from (5.11)) we deduce

$$\sum_{i=2}^{N-1} \alpha_i r_i \int_{-1}^{1} g(\bar{t}_i - y)(\bar{u}_N(\bar{t}_i) - \bar{u}_N(y)) dy - \sum_{i=2}^{N-1} \alpha_i r_{i-1} \int_{-1}^{1} g(\bar{t}_{i-1} - y)(\bar{u}_N(\bar{t}_{i-1}) - \bar{u}_N(y)) dy$$

$$= (-1)^{\frac{N}{2} + 1} \sum_{i=2}^{N-1} \alpha_i (2\alpha_i - \alpha_{i-1} - \alpha_{i+1}) \int_{-1}^{1} g(y) \bar{u}_N(y) dy$$

$$+ \frac{1}{N} \sum_{i=2}^{N-1} \alpha_i (\alpha_{i+1} + \alpha_{i-1} - 2\alpha_i) \int_{-1}^{1} g(y) dy$$

$$= 2(-1)^{\frac{N}{2} + 1} \int_{-1}^{1} g(y) \bar{u}_N(y) dy \left(\sum_{i=2}^{N-1} \alpha_i^2 - \sum_{i=2}^{N-1} \alpha_i \alpha_{i+1} \right)$$

$$+ \frac{2}{N} \int_{-1}^{1} g(y) dy \left(\sum_{i=2}^{N-1} \alpha_i \alpha_{i+1} - \sum_{i=2}^{N-1} \alpha_i^2 \right), \tag{6.17}$$

where we have used $\alpha_1 = \alpha_N = 0$. Setting (6.16) and (6.17) into (6.15), we obtain

$$\frac{1}{8} \bar{F}_r''(0) = \frac{2}{N} \int_{-1}^{1} g(y) dy \left(\sum_{i=2}^{N-1} \alpha_i^2 + \sum_{i=2}^{N-1} \alpha_i \alpha_{i+1} \right) - 2 \sum_{i=2}^{N-1} \sum_{k=2}^{N-1} \alpha_i \alpha_k \int_{\bar{t}_{i-1}}^{\bar{t}_i} \int_{\bar{t}_{k-1}}^{\bar{t}_k} g(x - y) dy dx$$

$$+ 2(-1)^{\frac{N}{2}+1} \int_{-1}^{1} g(y)\bar{u}_N(y)dy \left(\sum_{i=2}^{N-1} \alpha_i^2 - \sum_{i=2}^{N-1} \alpha_i\alpha_{i+1} \right)$$

and since $\alpha_1 = \alpha_N = 0$, this implies

$$\frac{1}{8}\bar{F}_r''(0) = \frac{2}{N} \int_{-1}^{1} g(y)dy \left(\sum_{i=1}^{N} \alpha_i^2 + \sum_{i=1}^{N-1} \alpha_i\alpha_{i+1} + \alpha_N\alpha_1 \right)$$
$$- 2 \sum_{i=1}^{N} \sum_{k=1}^{N} \alpha_i\alpha_k \int_{\bar{t}_{i-1}}^{\bar{t}_i} \int_{\bar{t}_{k-1}}^{\bar{t}_k} g(x-y)dydx$$
$$+ 2(-1)^{\frac{N}{2}+1} \int_{-1}^{1} g(y)\bar{u}_N(y)dy \left(\sum_{i=1}^{N} \alpha_i^2 - \sum_{i=1}^{N-1} \alpha_i\alpha_{i+1} - \alpha_N\alpha_1 \right)$$

which we can rewrite in matrix form as follows:

$$\frac{1}{8}\bar{F}_r''(0) = \frac{2}{N}\alpha^t A_N \alpha,$$

where

$$\alpha = (\alpha_1, \alpha_2, \ldots, \alpha_{N-1}, \alpha_N) \in \mathbb{R}^N, \quad A_N \in \mathbb{R}^{N \times N}$$

with A_N given by

$$A_N = B_N + C_N + D_N \tag{6.18}$$

with $B_N = (b_{ij}^{(N)})_{1 \leq i,j \leq N} \in \mathbb{R}^{N \times N}$, $D_N = (d_{ij}^{(N)})_{1 \leq i,j \leq N} \in \mathbb{R}^{N \times N}$ defined by

$$B_N = \int_{-1}^{1} g(y)dy \cdot \begin{bmatrix} 1 & \frac{1}{2} & 0 & \ldots & 0 & \frac{1}{2} \\ \frac{1}{2} & 1 & \ddots & \ddots & & 0 \\ 0 & \ddots & \ddots & \ddots & \ddots & \vdots \\ \vdots & \ddots & \ddots & \ddots & \ddots & 0 \\ 0 & & \ddots & \ddots & 1 & \frac{1}{2} \\ \frac{1}{2} & 0 & \ldots & 0 & \frac{1}{2} & 1 \end{bmatrix} \in \mathbb{R}^{N \times N}, \tag{6.19}$$

$$D_N = N(-1)^{\frac{N}{2}+1} \int_{-1}^{1} g(y)\bar{u}_N(y)dy \cdot \begin{bmatrix} 1 & -\frac{1}{2} & 0 & \ldots & 0 & -\frac{1}{2} \\ -\frac{1}{2} & 1 & \ddots & \ddots & & 0 \\ 0 & \ddots & \ddots & \ddots & \ddots & \vdots \\ \vdots & \ddots & \ddots & \ddots & \ddots & 0 \\ 0 & & \ddots & \ddots & 1 & -\frac{1}{2} \\ -\frac{1}{2} & 0 & \ldots & 0 & -\frac{1}{2} & 1 \end{bmatrix} \in \mathbb{R}^{N \times N} \tag{6.20}$$

and $C_N = (c_{ij}^{(N)})_{1 \leq i,j \leq N} \in \mathbb{R}^{N \times N}$ given by

$$c_{ij}^{(N)} = -N \int_{\bar{t}_{i-1}}^{\bar{t}_i} \int_{\bar{t}_{j-1}}^{\bar{t}_j} g(x-y)dydx = -N \int_{-1+\frac{2}{N}(i-1)}^{-1+\frac{2}{N}i} \int_{-1+\frac{2}{N}(j-1)}^{-1+\frac{2}{N}j} g(x-y)dydx$$

$$= -N \int_{\frac{2}{N}(i-1)}^{\frac{2}{N}i} \int_{\frac{2}{N}(j-1)}^{\frac{2}{N}j} g(x-y)\,dy\,dx$$

for $i,j = 1,\dots,N$. Clearly, $B_N = (b_{|i-j|}^{(N)})_{i,j=1,\dots,N}$, $D_N = (d_{|i-j|}^{(N)})_{i,j=1,\dots,N}$ are symmetric with coefficients given by (6.11), and both matrices are circulant. As for C_N, we have

$$c_{i+1,j+1}^{(N)} = -N \int_{\frac{2}{N}i}^{\frac{2}{N}(i+1)} \int_{\frac{2}{N}j}^{\frac{2}{N}(j+1)} g(x-y)\,dy\,dx = -N \int_{\frac{2}{N}(i-1)}^{\frac{2}{N}i} \int_{\frac{2}{N}(j-1)}^{\frac{2}{N}j} g(x-y)\,dy\,dx = c_{ij}^{(N)}$$

for $i,j = 1,\dots,N-1$. Furthermore, C_N is symmetric which follows from the symmetry of g:

$$c_{ji}^{(N)} = -N \int_{\bar{t}_{j-1}}^{\bar{t}_j} \int_{\bar{t}_{i-1}}^{\bar{t}_i} g(x-y)\,dy\,dx = -N \int_{\bar{t}_{i-1}}^{\bar{t}_i} \int_{\bar{t}_{j-1}}^{\bar{t}_j} g(x-y)\,dx\,dy =$$

$$= -N \int_{\bar{t}_{i-1}}^{\bar{t}_i} \int_{\bar{t}_{j-1}}^{\bar{t}_j} g(y-x)\,dx\,dy = -N \int_{\bar{t}_{i-1}}^{\bar{t}_i} \int_{\bar{t}_{j-1}}^{\bar{t}_j} g(x-y)\,dy\,dx = c_{ij}^{(N)}.$$

Thus, $c_{ij}^{(N)} = c_{|i-j|}^{(N)}$ for $i,j = 1,\dots,N$, where $c_k^{(N)}$ are given by

$$c_k^{(N)} = c_{1,k+1}^{(N)} = -N \int_0^{\frac{2}{N}} \int_{\frac{2}{N}k}^{\frac{2}{N}(k+1)} g(x-y)\,dy\,dx$$

for $k = 0,\dots,N-1$. Using the 2-periodicity and the symmetry of g, we infer

$$c_{N-k}^{(N)} = -N \int_0^{\frac{2}{N}} \int_{\frac{2}{N}(N-k)}^{\frac{2}{N}(N-k+1)} g(x-y)\,dy\,dx = -N \int_0^{\frac{2}{N}} \int_{-\frac{2}{N}k}^{-\frac{2}{N}(k-1)} g(x-y)\,dy\,dx =$$

$$= -N \int_0^{\frac{2}{N}} \int_{\frac{2}{N}(k-1)}^{\frac{2}{N}k} g(x+y)\,dy\,dx = -N \int_{-\frac{2}{N}}^0 \int_{\frac{2}{N}(k-1)}^{\frac{2}{N}k} g(y-x)\,dy\,dx =$$

$$= -N \int_0^{\frac{2}{N}} \int_{\frac{2}{N}k}^{\frac{2}{N}(k+1)} g(y-x)\,dy\,dx = -N \int_0^{\frac{2}{N}} \int_{\frac{2}{N}k}^{\frac{2}{N}(k+1)} g(x-y)\,dy\,dx = c_k^{(N)}$$

for $k = 1,\dots,N-1$, which implies the circulant structure of C_N. The rest of the Theorem now easily follows. $\square$

Chapter 7

Circulant Matrices

7.1 Eigenvalues of Circulant Matrices

To show that the function $\bar{F}_r$ has a local minimum at 0 for every $r = (r_1, \ldots, r_{N-1}) \in \mathbb{R}^{N-1}$ satisfying the periodicity condition (6.8), we consider the eigenvalues of the matrix A_N in Theorem 6.4. It has been pointed out by Varga [46] in 1954 already that the eigenvalues of circulant matrices can be obtained explicitly by a discrete Fourier transform of the coefficients. Indeed, the special structure of A_N allows us to write down the eigenvalues and corresponding eigenvectors. We cite the following result (see Theorem 3.2.2 in [16], Chapter 3 in [25]), which we will prove for the sake of completeness.

Theorem 7.1 *Let $M \in \mathbb{C}^{n \times n}$ be a circulant matrix, i.e.*

$$
M = \begin{bmatrix}
\alpha_0 & \alpha_1 & \alpha_2 & \cdots & & \alpha_{n-1} \\
\alpha_{n-1} & \alpha_0 & \alpha_1 & \alpha_2 & & \vdots \\
\alpha_{n-2} & \alpha_{n-1} & \alpha_0 & \alpha_1 & \ddots & \vdots \\
\vdots & \ddots & \ddots & \ddots & & \alpha_2 \\
\vdots & & \ddots & \ddots & \ddots & \alpha_1 \\
\alpha_1 & \cdots & & \alpha_{n-2} & \alpha_{n-1} & \alpha_0
\end{bmatrix},
$$

or, equivalently, $M = (\alpha_{i-j})_{i,j=1,\ldots,n}$ with

$$
\alpha_{-k} = \alpha_{n-k} \quad for \quad k = 1, \ldots, n-1.
$$

Then the eigenvalues $\lambda_0, \ldots, \lambda_{n-1} \in \mathbb{C}$ of M are given by

$$
\lambda_t = \sum_{k=0}^{n-1} \alpha_k \exp\left[-\frac{2\pi i k t}{n}\right]
$$

with corresponding eigenvectors $v_t = (v_t^{(0)}, \ldots, v_t^{(n-1)}) \in \mathbb{C}^n$ for $t = 0, \ldots, n-1$, where

$$
v_t^{(k)} = \exp\left[\frac{2\pi i k t}{n}\right] \quad for \quad k = 0, \ldots, n-1
$$

for $t = 0, \ldots, n-1$.

Proof: Let $M \in \mathbb{C}^{n \times n}$, $v_t = (v_t^{(0)}, \ldots, v_t^{(n-1)}) \in \mathbb{C}^n$ $(t = 0, \ldots, n-1)$ as given in the theorem. Set $Mv_t = ((Mv_t)_1, \ldots, (Mv_t)_n)$. Then for $l = 1, \ldots, n$ we have

$$
\begin{aligned}
(Mv_t)_l &= \sum_{j=1}^{n} \alpha_{l-j} v_t^{(j-1)} = \sum_{j=1}^{l} \alpha_{l-j} v_t^{(j-1)} + \sum_{j=l+1}^{n} \alpha_{-(j-l)} v_t^{(j-1)} \\
&= \sum_{j=1}^{l} \alpha_{l-j} v_t^{(j-1)} + \sum_{j=l+1}^{n} \alpha_{n-j+l} v_t^{(j-1)} = \sum_{k=0}^{l-1} \alpha_k v_t^{(l-k-1)} + \sum_{k=l}^{n-1} \alpha_k v_t^{(n-k+l-1)} \\
&= \sum_{k=0}^{l-1} \alpha_k \exp\left[\frac{2\pi(l-k-1)ti}{n}\right] + \sum_{k=l}^{n-1} \alpha_k \exp\left[\frac{2\pi(n-k+l-1)ti}{n}\right] \\
&= \sum_{k=0}^{n-1} \alpha_k \exp\left[\frac{2\pi(l-k-1)ti}{n}\right] = \sum_{k=0}^{n-1} \alpha_k \exp\left[-\frac{2\pi kti}{n}\right] \exp\left[\frac{2\pi(l-1)ti}{n}\right] \\
&= \lambda_t v_t^{(l-1)},
\end{aligned}
$$

so that $Mv_t = ((Mv_t)_1, \ldots, (Mv_t)_n) = \lambda_t(v_t^{(0)}, \ldots, v_t^{(n-1)}) = \lambda_t v_t$ for $t = 0, \ldots, n-1$. Since $v_t \neq 0$ due to $v_t^{(0)} = 1$, λ_t is an eigenvalue with eigenvector v_t. To make sure that these are indeed all eigenvalues, it is sufficient to show that $v_s \perp v_t$ for $s \neq t$. Let s, $t \in \{0, \ldots, n-1\}$ with $t > s$. Since $\frac{t-s}{n} \notin \mathbb{Z}$, we have $\exp\left[\frac{2\pi(t-s)i}{n}\right] \neq 1$, so that

$$
\begin{aligned}
(v_t, v_s) &= \sum_{k=0}^{n-1} v_t^{(k)} \overline{v_s^{(k)}} = \sum_{k=0}^{n-1} \exp\left[\frac{2\pi k(t-s)i}{n}\right] = \sum_{k=0}^{n-1} \exp\left[\frac{2\pi(t-s)i}{n}\right]^k \\
&= \frac{1 - \exp\left[\frac{2\pi(t-s)i}{n}\right]^n}{1 - \exp\left[\frac{2\pi(t-s)i}{n}\right]} = \frac{1 - \exp[2\pi(t-s)i]}{1 - \exp\left[\frac{2\pi(t-s)i}{n}\right]} = 0
\end{aligned}
$$

due to the Euler formula. Theorem 7.1 now follows. $\square$

Transferring this to the special case of a symmetric matrix $M \in \mathbb{R}^{n \times n}$, where n is even, we obtain (also see [19])

Theorem 7.2 *Let $n \in \mathbb{N}$ even, and $M \in \mathbb{R}^{n \times n}$ circulant and symmetric, i.e.*

$$
M = \begin{bmatrix}
\alpha_0 & \alpha_1 & \alpha_2 & \cdots & \alpha_2 & \alpha_1 \\
\alpha_1 & \alpha_0 & \alpha_1 & \alpha_2 & & \alpha_2 \\
\alpha_2 & \alpha_1 & \alpha_0 & \alpha_1 & \ddots & \vdots \\
\vdots & \ddots & \ddots & \ddots & & \alpha_2 \\
\vdots & & \ddots & \ddots & \ddots & \alpha_1 \\
\alpha_1 & \cdots & \cdots & \alpha_2 & \alpha_1 & \alpha_0,
\end{bmatrix},
$$

or, equivalently, $M = (\alpha_{|i-j|})_{i,j=1,\ldots,n}$ with

$$
\alpha_{n-k} = \alpha_k \quad \text{for} \quad k = 1, \ldots, n-1.
$$

Then the eigenvalues $\lambda_0, \ldots, \lambda_{\frac{n}{2}} \in \mathbb{R}$ of M are given by

$$\lambda_t = \alpha_0 + 2 \sum_{k=1}^{\frac{n}{2}-1} \alpha_k \cos\left[\frac{2\pi kt}{n}\right] + (-1)^t \alpha_{\frac{n}{2}}$$

for $t = 0, \ldots, \frac{n}{2}$, where $\lambda_1, \ldots, \lambda_{\frac{n}{2}-1}$ are double. Corresponding eigenvectors are given by $v_t = (v_t^{(0)}, \ldots, v_t^{(n-1)})$, where

$$v_t^{(k)} = \cos\left[\frac{2\pi kt}{n}\right] \quad for \quad k = 0, \ldots, n-1$$

for $t = 0, \ldots, \frac{n}{2}$.

Proof: Since M is symmetric, M has n real eigenvalues which by Theorem 7.1 are given by

$$\lambda_t = \sum_{k=0}^{n-1} \alpha_k \exp\left[-\frac{2\pi ikt}{n}\right] = \sum_{k=0}^{n-1} \alpha_k \cos\left[\frac{2\pi kt}{n}\right]$$

for $t = 0, \ldots, n-1$. For $t = 1, \ldots, \frac{n}{2} - 1$, we have

$$\lambda_{n-t} = \sum_{k=0}^{n-1} \alpha_k \cos\left[\frac{2\pi k(n-t)}{n}\right] = \sum_{k=0}^{n-1} \alpha_k \cos\left[2\pi k - \frac{2\pi kt}{n}\right] = \sum_{k=0}^{n-1} \alpha_k \cos\left[\frac{2\pi kt}{n}\right] = \lambda_t,$$

which implies that $\lambda_1, \ldots, \lambda_{\frac{n}{2}-1}$ are double, so that we only have to consider the indices $t = 0, \ldots, \frac{n}{2}$. Since $\alpha_k = \alpha_{n-k}$ and $\cos[\frac{2\pi kt}{n}] = \cos[\frac{2\pi(n-k)t}{n}]$, we obtain

$$\lambda_t = \sum_{k=0}^{n-1} \alpha_k \cos\left[\frac{2\pi kt}{n}\right] = \sum_{k=0}^{\frac{n}{2}-1} \alpha_k \cos\left[\frac{2\pi kt}{n}\right] + \sum_{k=\frac{n}{2}}^{n-1} \alpha_{n-k} \cos\left[\frac{2\pi(n-k)t}{n}\right]$$

$$= \sum_{k=0}^{\frac{n}{2}-1} \alpha_k \cos\left[\frac{2\pi kt}{n}\right] + \sum_{k=1}^{\frac{n}{2}} \alpha_k \cos\left[\frac{2\pi kt}{n}\right] = \alpha_0 + 2 \sum_{k=1}^{\frac{n}{2}-1} \alpha_k \cos\left[\frac{2\pi kt}{n}\right] + (-1)^t \alpha_{\frac{n}{2}},$$

where we used $\cos(\pi t) = (-1)^t$. Since M is symmetric, eigenvectors are given by the real and imaginary parts of the complex eigenvectors given in Theorem 7.1, and the proof is complete. $\qquad\square$

Remark: Note that the eigenspaces of a circulant matrix M do only depend on the dimension, not on (the coefficients of) the matrix itself.

7.2   Application to A_N

For the matrix A_N defined in Theorem 6.4, we immediately deduce that all eigenvalues are given by $\lambda_0^{(N)}, \ldots, \lambda_{\frac{N}{2}}^{(N)}$, where

$$\lambda_t^{(N)} = a_0^{(N)} + (-1)^t a_{\frac{N}{2}}^{(N)} + 2 \sum_{k=1}^{\frac{N}{2}-1} a_k^{(N)} \cos\left[\frac{2\pi kt}{N}\right] \quad \text{for} \quad t = 0, \ldots, \frac{N}{2}, \qquad (7.1)$$

the coefficients given by (6.14), (6.12), (6.11). $\lambda_1^{(N)}, \ldots, \lambda_{\frac{N}{2}-1}^{(N)}$ are double eigenvalues, and for every $t \in \left\{0, \ldots, \frac{N}{2}\right\}$, an eigenvector corresponding to $\lambda_t^{(N)}$ is given by

$$v_t = \left(v_t^{(0)}, \ldots, v_t^{(N-1)}\right) \in \mathbb{R}^N,$$
$$v_t^{(k)} = \cos\left[\frac{2\pi kt}{N}\right], \quad k = 0, \ldots, N-1. \qquad (7.2)$$

7.2.1   The Eigenvalue Zero

First of all, we will deal with the "outer" eigenvalues $\lambda_0^{(N)}$ and $\lambda_{\frac{N}{2}}^{(N)}$. These are characterized in the following theorem:

Theorem 7.3 *For every even number $N \in \mathbb{N}$, let A_N as given in Theorem 6.4 with eigenvalues $\lambda_t^{(N)}$, $t = 0, \ldots, \frac{N}{2}$, given by (7.1). Then*

$$\lambda_0^{(N)} = \lambda_{\frac{N}{2}}^{(N)} = 0. \qquad (7.3)$$

Proof: Recall (6.14), (6.12) and (6.11). Using (7.1) and the 2-periodicity and symmetry of g, we compute

$$\lambda_0^{(N)} = a_0^{(N)} + a_{\frac{N}{2}}^{(N)} + 2 \sum_{k=1}^{\frac{N}{2}-1} a_k^{(N)} = b_0^{(N)} + d_0^{(N)} + c_0^{(N)} + c_{\frac{N}{2}}^{(N)} + 2b_1^{(N)} + 2d_1^{(N)} + 2 \sum_{k=1}^{\frac{N}{2}-1} c_k^{(N)}$$

$$= 2b_0^{(N)} + c_0^{(N)} + c_{\frac{N}{2}}^{(N)} + 2 \sum_{k=1}^{\frac{N}{2}-1} c_k^{(N)} = 2b_0^{(N)} + \sum_{k=0}^{\frac{N}{2}-1} c_k^{(N)} + \sum_{k=1}^{\frac{N}{2}} c_k^{(N)}$$

$$= 2 \int_0^2 g(y)dy - N \sum_{k=0}^{\frac{N}{2}-1} \int_0^{\frac{2}{N}} \int_{\frac{2}{N}k}^{\frac{2}{N}(k+1)} g(x-y)dydx$$

$$- N \sum_{k=1}^{\frac{N}{2}} \int_0^{\frac{2}{N}} \int_{\frac{2}{N}k}^{\frac{2}{N}(k+1)} g(x-y)dydx$$

$$=2\int_0^2 g(y)dy - N\int_0^{\frac{2}{N}}\int_0^1 g(x-y)dydx - N\int_0^{\frac{2}{N}}\int_{\frac{2}{N}}^{1+\frac{2}{N}} g(x-y)dydx$$

$$=2\int_0^2 g(y)dy - N\int_0^{\frac{2}{N}}\int_0^1 g(x-y)dydx - N\int_{-\frac{2}{N}}^0\int_0^1 g(x-y)dydx$$

$$=2\int_0^2 g(y)dy - N\int_0^{\frac{2}{N}}\int_0^1 g(x-y)dydx - N\int_0^{\frac{2}{N}}\int_0^1 g(-x-y)dydx$$

$$=2\int_0^2 g(y)dy - N\int_0^{\frac{2}{N}}\int_0^1 g(y-x)dydx - N\int_0^{\frac{2}{N}}\int_{-1}^0 g(y-x)dydx$$

$$=2\int_0^2 g(y)dy - N\int_0^{\frac{2}{N}}\int_{-1}^1 g(y-x)dydx = 2\int_0^2 g(y)dy - N\int_0^{\frac{2}{N}}\int_{-1}^1 g(y)dydx$$

$$=2\int_0^2 g(y)dy - N\cdot\frac{2}{N}\int_{-1}^1 g(y)dy = 0,$$

which implies the first equation in the theorem. To calculate $\lambda_{\frac{N}{2}}^{(N)}$, one has to work harder. Due to (7.1), we have

$$\lambda_{\frac{N}{2}}^{(N)} = a_0^{(N)} + (-1)^{\frac{N}{2}}a_{\frac{N}{2}}^{(N)} + 2\sum_{k=1}^{\frac{N}{2}-1}(-1)^k a_k^{(N)}$$

$$= b_0^{(N)} + c_0^{(N)} + d_0^{(N)} + (-1)^{\frac{N}{2}}c_{\frac{N}{2}}^{(N)} - 2b_1^{(N)} - 2d_1^{(N)} + 2\sum_{k=1}^{\frac{N}{2}-1}(-1)^k c_k^{(N)}$$

$$= 2d_0^{(N)} + c_0^{(N)} + (-1)^{\frac{N}{2}}c_{\frac{N}{2}}^{(N)} + 2\sum_{k=1}^{\frac{N}{2}-1}(-1)^k c_k^{(N)}. \tag{7.4}$$

To calculate $d_0^{(N)}$ and $c_k^{(N)}$, we choose a function $H \in C^2[0,1]$ satisfying $H'' = g$ on $[0,1]$. Recalling (6.1), we have $\frac{2}{N}k = -1 + \frac{2}{N}(k + \frac{N}{2}) = \bar{t}_{k+\frac{N}{2}}$ for $k = 0,\ldots,\frac{N}{2}$, so that due to Lemma 6.1 (i), $\bar{u}'_N(x) = (-1)^{k+\frac{N}{2}+1}$ for every $x \in (\frac{2}{N}(k-1), \frac{2}{N}k)$, $k = 1,\ldots,\frac{N}{2}$. Using the symmetry of both g and $\bar{u}_N$, we obtain

$$\int_{-1}^1 g(y)\bar{u}_N(y)dy = 2\int_0^1 g(y)\bar{u}_N(y)dy = 2\int_0^1 H''(y)\bar{u}_N(y)dy$$

$$= -2\int_0^1 H'(y)\bar{u}'_N(y)dy + 2(H'(1)\bar{u}_N(1) - H'(0)\bar{u}_N(0))$$

$$= -2\sum_{k=1}^{\frac{N}{2}}\int_{\frac{2}{N}(k-1)}^{\frac{2}{N}k} H'(y)\bar{u}'_N(y)dy + 2(H'(1)\bar{u}_N(1) - H'(0)\bar{u}_N(0))$$

$$= -2\sum_{k=1}^{\frac{N}{2}}(-1)^{k+\frac{N}{2}+1}\int_{\frac{2}{N}(k-1)}^{\frac{2}{N}k} H'(y)dy + 2(H'(1)\bar{u}_N(1) - H'(0)\bar{u}_N(0))$$

$$=2\sum_{k=1}^{\frac{N}{2}}(-1)^{k+\frac{N}{2}}[H(\frac{2}{N}k)-H(\frac{2}{N}(k-1))]+2(H'(1)\bar{u}_N(1)-H'(0)\bar{u}_N(0))$$

$$=2(-1)^{\frac{N}{2}}\left[\sum_{k=1}^{\frac{N}{2}}(-1)^kH(\frac{2}{N}k)+\sum_{k=0}^{\frac{N}{2}-1}(-1)^kH(\frac{2}{N}k)\right]$$
$$+2(H'(1)\bar{u}_N(1)-H'(0)\bar{u}_N(0))$$

$$=4(-1)^{\frac{N}{2}}\sum_{k=1}^{\frac{N}{2}-1}(-1)^kH(\frac{2}{N}k)+2(-1)^{\frac{N}{2}}H(0)+2H(1)$$
$$+2(H'(1)\bar{u}_N(1)-H'(0)\bar{u}_N(0)).$$

By Lemma 6.1 (ii) and (viii), we have $\bar{u}_N(1)=-\frac{1}{N}$ and $\bar{u}_N(0)=\bar{u}_N(\bar{t}_{\frac{N}{2}})=\frac{1}{N}(-1)^{\frac{N}{2}+1}$, which implies

$$\int_{-1}^{1}g(y)\bar{u}_N(y)dy=4(-1)^{\frac{N}{2}}\sum_{k=1}^{\frac{N}{2}-1}(-1)^kH(\frac{2}{N}k)+2(-1)^{\frac{N}{2}}H(0)+2H(1)$$
$$-\frac{2}{N}H'(1)+\frac{2}{N}(-1)^{\frac{N}{2}}H'(0)$$

hence

$$2d_0^{(N)}=2N(-1)^{\frac{N}{2}+1}\int_{-1}^{1}g(y)\bar{u}_N(y)dy$$

$$=-8N\sum_{k=1}^{\frac{N}{2}-1}(-1)^kH(\frac{2}{N}k)-4NH(0)$$

$$-4N(-1)^{\frac{N}{2}}H(1)+4(-1)^{\frac{N}{2}}H'(1)-4H'(0). \tag{7.5}$$

To compute the other terms in (7.4), let $k\in\{1,\dots,\frac{N}{2}-1\}$. Then for $x\in(\frac{2}{N}k,\frac{2}{N}(k+1))$ and $y\in(0,\frac{2}{N})$, we have $x-y\in(0,1)$, so that $g(x-y)=H''(x-y)$ holds, and we obtain

$$c_k^{(N)}=-N\int_0^{\frac{2}{N}}\int_{\frac{2}{N}k}^{\frac{2}{N}(k+1)}g(x-y)dxdy=-N\int_0^{\frac{2}{N}}\int_{\frac{2}{N}k}^{\frac{2}{N}(k+1)}H''(x-y)dxdy$$

$$=-N\int_0^{\frac{2}{N}}[H'(\frac{2}{N}(k+1)-y)-H'(\frac{2}{N}k-y)]dy$$

$$=N\left[2H(\frac{2}{N}k)-H(\frac{2}{N}(k+1))-H(\frac{2}{N}(k-1))\right],$$

Calculating $c_0^{(N)}$ requires more caution, as $x-y$ does not lie in $(0,1)$ for all $x,y\in(0,\frac{2}{N})$ so that we will have to use the symmetry of g first to use $H''=g$, which works as follows

$$c_0^{(N)}=-N\int_0^{\frac{2}{N}}\int_0^{\frac{2}{N}}g(x-y)dxdy$$

$$= -N \int_0^{\frac{2}{N}} \int_0^y g(y-x)dxdy - N \int_0^{\frac{2}{N}} \int_y^{\frac{2}{N}} g(x-y)dxdy$$

$$= -N \int_0^{\frac{2}{N}} \int_0^y H''(y-x)dxdy - N \int_0^{\frac{2}{N}} \int_y^{\frac{2}{N}} H''(x-y)dxdy$$

$$= N \int_0^{\frac{2}{N}} [H'(0) - H'(y)]dy - N \int_0^{\frac{2}{N}} [H'(\frac{2}{N} - y) - H'(0)]dy$$

$$= 2H'(0) - NH(y)\Big|_0^{\frac{2}{N}} + NH(\frac{2}{N} - y)\Big|_0^{\frac{2}{N}} + 2H'(0)$$

$$= 4H'(0) - 2NH(\frac{2}{N}) + 2NH(0).$$

In order to compute $c_{\frac{N}{2}}^{(N)}$, we first observe that $g(-z+1) = g(-z-1) = g(z+1)$ for every $z \in \mathbb{R}$ (which follows from the symmetry and the 2-periodicity of g). Thus

$$c_{\frac{N}{2}}^{(N)} = -N \int_0^{\frac{2}{N}} \int_1^{1+\frac{2}{N}} g(x-y)dxdy = -N \int_0^{\frac{2}{N}} \int_0^{\frac{2}{N}} g(x-y+1)dxdy$$

$$= -N \int_0^{\frac{2}{N}} \int_0^y g(x-y+1)dxdy - N \int_0^{\frac{2}{N}} \int_y^{\frac{2}{N}} g(y-x+1)dxdy.$$

As for the left-hand integral, $x - y + 1$ obviously lies in $(0,1)$, as does $y - x + 1$ on the domain of the right-hand integral, so that we can use $H'' = g$ to obtain

$$c_{\frac{N}{2}}^{(N)} = -N \int_0^{\frac{2}{N}} \int_0^y H''(x-y+1)dxdy - N \int_0^{\frac{2}{N}} \int_y^{\frac{2}{N}} H''(y-x+1)dxdy$$

$$= -N \int_0^{\frac{2}{N}} [H'(1) - H'(1-y)]dy + N \int_0^{\frac{2}{N}} [H'(y - \frac{2}{N} + 1) - H'(1)]dx$$

$$= -2H'(1) - NH(1-y)\Big|_0^{\frac{2}{N}} + NH(y - \frac{2}{N} + 1)\Big|_0^{\frac{2}{N}} - 2H'(1)$$

$$= -4H'(1) - 2NH(1 - \frac{2}{N}) + 2NH(1).$$

Combining these formulas, we deduce

$$c_0^{(N)} + (-1)^{\frac{N}{2}} c_{\frac{N}{2}}^{(N)} + 2 \sum_{k=1}^{\frac{N}{2}-1} (-1)^k c_k^{(N)} =$$

$$= 4H'(0) - 2NH(\frac{2}{N}) + 2NH(0) + 4(-1)^{\frac{N}{2}+1}H'(1) + 2N(-1)^{\frac{N}{2}+1}H(1 - \frac{2}{N})$$

$$+ 2N(-1)^{\frac{N}{2}}H(1) + 2N \sum_{k=1}^{\frac{N}{2}-1} (-1)^k \left[2H(\frac{2}{N}k) - H(\frac{2}{N}(k+1)) - H(\frac{2}{N}(k-1)) \right]$$

$$= 4H'(0) - 2NH(\frac{2}{N}) + 2NH(0) - 4(-1)^{\frac{N}{2}}H'(1) + 2N(-1)^{\frac{N}{2}+1}H(1 - \frac{2}{N})$$

$$+ 2N(-1)^{\frac{N}{2}}H(1) + 4N \sum_{k=1}^{\frac{N}{2}-1} (-1)^k H(\frac{2}{N}k) + 2N \sum_{k=2}^{\frac{N}{2}} (-1)^k H(\frac{2}{N}k)$$

$$+ 2N \sum_{k=0}^{\frac{N}{2}-2} (-1)^k H(\frac{2}{N}k)$$

$$= 4H'(0) - 2NH(\frac{2}{N}) + 2NH(0) - 4(-1)^{\frac{N}{2}} H'(1) + 2N(-1)^{\frac{N}{2}+1} H(1 - \frac{2}{N})$$

$$+ 2N(-1)^{\frac{N}{2}} H(1) + 8N \sum_{k=1}^{\frac{N}{2}-1} (-1)^k H(\frac{2}{N}k) + 2NH(\frac{2}{N}) + 2N(-1)^{\frac{N}{2}} H(1)$$

$$+ 2NH(0) + 2N(-1)^{\frac{N}{2}} H(1 - \frac{2}{N})$$

$$= 4H'(0) + 4NH(0) - 4(-1)^{\frac{N}{2}} H'(1) + 4N(-1)^{\frac{N}{2}} H(1) + 8N \sum_{k=1}^{\frac{N}{2}-1} (-1)^k H(\frac{2}{N}k)$$

$$= -2d_0^{(N)}$$

because of (7.5). $\lambda_{\frac{N}{2}}^{(N)} = 0$ now immediately follows from (7.4). $\qquad\square$

7.2.2   A Positivity Condition for A_N

From Theorem 6.4, it follows that the local minimality of $\bar{u}_N$ under any perturbation of $\bar{u}_N$ that keeps the periodicity condition is connected to the positive definiteness of the matrix A_N on the $N-2$-dimensional subspace $X_N \subset \mathbb{R}^N$ given by

$$X_N = \{\alpha = (\alpha_1, \ldots, \alpha_N) \in \mathbb{R}^N \,|\, \alpha_1 = \alpha_N = 0\}. \tag{7.6}$$

From Theorem 7.3, we know that 0 is an eigenvalue of A_N with multiplicity ≥ 2. To get positive definiteness, it is thus necessary that the double eigenvalues $\lambda_t^{(N)}$, $t = 1, \ldots, \frac{N}{2} - 1$ are positive. The following theorem shows that this condition is also sufficient:

Theorem 7.4 *For an even number $N \in \mathbb{N}$, let A_N as given in Theorem 6.4 with eigenvalues $\lambda_t^{(N)}$, $t = 0, \ldots, \frac{N}{2}$, given by (7.1). Let X_N be the $N-2$-dimensional subspace of $\mathbb{R}^N$ as given by (7.6). Assume that*

$$\lambda_t^{(N)} > 0 \quad \text{for} \quad t = 1, \ldots, \frac{N}{2} - 1.$$

Then

$$\alpha^t A_N \alpha > 0 \quad \text{for every} \quad \alpha \in X_N \setminus \{0\}, \tag{7.7}$$

and for every $r = (r_1, \ldots, r_{N-1}) \in \mathbb{R}^{N-1}$ satisfying the periodicity condition (6.8), $\bar{F}_r$ as defined by (6.9) satisfies

$$\bar{F}_r'(0) = 0, \quad \bar{F}_r''(0) > 0$$

and thus has a local minimum at 0.

Proof: Since for every $t = 1, \ldots, \frac{N}{2} - 1$, $\lambda_t^{(N)} > 0$ is a double eigenvalue and $\lambda_0^{(N)} = \lambda_{\frac{N}{2}}^{(N)} = 0$ due to Theorem 7.3, the matrix A_N has rank $N - 2$. Thus $\ker A_N$, which is the eigenspace of $\lambda_0^{(N)} = \lambda_{\frac{N}{2}}^{(N)} = 0$, due to (7.2) is given by

$$\ker A_N = \text{span}[v_0^{(N)}, v_{\frac{N}{2}}^{(N)}] = \text{span}[(1, 1, \ldots, 1), (1, -1, 1, \ldots, -1)]$$

(note that N is even), and $\mathbb{R}^N$ is decomposed as follows:

$$\mathbb{R}^N = X_N \oplus \ker A_N.$$

Indeed, $\dim X_N + \dim \ker A_N = N - 2 + 2 = N$, and any $\alpha \in X_N \cap \ker A_N$ satisfies

$$(0, \alpha_2, \ldots, \alpha_{N-1}, 0) = \alpha = (\lambda + \mu, \lambda - \mu, \lambda + \mu, \ldots, \lambda - \mu)$$

for certain λ, $\mu \in \mathbb{R}$. In particular, $\lambda + \mu = 0$, $\lambda - \mu = 0$, which implies $\lambda = \mu = 0$. Thus, $\alpha = 0$ and since α was chosen arbitrarily, we conclude $X_N \cap \ker A_N = \{0\}$. We immediately deduce

$$A_N \alpha \neq 0 \quad \text{for every} \quad \alpha \in X_N \setminus \{0\}.$$

Since A_N is symmetric and all eigenvalues of A_N are non-negative, A_N is positive semi-definite, so that we find a symmetric positive semi-definite matrix $B_N = A_N^{\frac{1}{2}} \in \mathbb{R}^{N \times N}$ with $B_N^2 = A_N$ and

$$B_N \alpha \neq 0 \quad \text{for every} \quad \alpha \in X_N \setminus \{0\}.$$

Now, let $\alpha \in X_N$ such that $\alpha^t A_N \alpha = 0$. Then

$$(B_N \alpha)^t (B_N \alpha) = \alpha^t B_N B_N \alpha = \alpha^t A_N \alpha = 0$$

which implies $B_N \alpha = 0$, and since $\alpha \in X_N$, we conclude $\alpha = 0$. Thus, $\alpha^t A_N \alpha \neq 0$ for every $\alpha \in X_N \setminus \{0\}$, which due to the positive semi-definiteness of A_N yields (7.7). Finally, let $r = (r_1, \ldots, r_{N-1}) \in \mathbb{R}^{(N-1)}$ satisfy (6.8). From Theorem 6.3, we get

$$\bar{F}_r'(0) = 0.$$

Let $\alpha \in \mathbb{R}^N$ as defined in (5.11), then Theorem 6.4 yields $\alpha \in X_N$ and

$$\bar{F}_r''(0) = \alpha^t A_N \alpha > 0,$$

so that $\bar{F}_r$ attains a local minimum at 0. $\qquad\square$

Chapter 8

Local Minimality Criteria

8.1 The Nonlocal Energy as a Function of the Corners

In the following, we will show that if the "r-variation" F_r of the equidistant sawtooth function $\bar{u}_N$ attains a local minimum at 0 for every $r \in \mathbb{R}^{N-1}$ satisfying the periodicity condition (6.8), then $\bar{u}_N$ is an H^1-local minimizer of E in the set of all $u \in \mathcal{S}^0_{per}(-1,1)$ with N corners in $[-1,1)$. As before, N is a fixed even number.

Theorem 7.7 then suggests to examine the eigenvalues $\lambda_1^{(N)}, \ldots, \lambda_{\frac{N}{2}-1}^{(N)}$ to obtain local minimizers of I^ε via Corollary 4.9. This correlation is resumed in Corollary 8.6.

We define the set of sawtooth functions on $(-1,1)$ with slope $u' \in \{\pm 1\}$ only and zero average by

$$\mathcal{S}^0(-1,1) = \mathcal{S}(-1,1) \cap \{\int_{-1}^1 u(x)dx = 0\}.$$

For a fixed $N \in \mathbb{N}$, we view the nonlocal energy E as a function of the inner corners $t_1, \ldots, t_{N-1}$ of arbitrary sawtooth functions $u \in \mathcal{S}^0(-1,1)$ with $\#(Su' \cap (-1,1)) = N-1$ (i.e. u has $N-1$ inner corners). To do so, we define an open set $\Delta_N \subset \mathbb{R}^{N-1}$ by

$$\Delta_N = \{t = (t_1, \ldots, t_{N-1}) \in \mathbb{R}^{N-1} | -1 < t_1 < \cdots < t_{N-1} < 1\}$$

and a function

$$\mathcal{U}_N : \Delta_N \to \mathcal{S}^0(-1,1),$$

where for every $t = (t_1, \ldots, t_{N-1}) \in \Delta_N$, we choose $\mathcal{U}_N(t)$ as the unique element of $\mathcal{S}^0(-1,1)$ satisfying

$$(\mathcal{U}_N(t))'(x) = (-1)^{i-1} \quad \text{for} \quad x \in (t_{i-1}, t_i), \quad i = 1, \ldots, N. \tag{8.1}$$

Here, $t_0 = -1$, $t_N = 1$ is set. We define $\mathcal{F}_N = E \circ \mathcal{U}_N : \Delta_N \to \mathbb{R}$ as the nonlocal energy of $\mathcal{U}_N(t)$, i.e.

$$\mathcal{F}_N(t) = E(\mathcal{U}_N(t)) \quad \text{for every} \quad t \in \Delta_N, \tag{8.2}$$

aiming to show that under certain conditions, $\mathcal{F}_N$ attains a local minimum at the equidistant vector $(\bar{t}_1, \ldots, \bar{t}_{N-1}) \in \Delta_N$ (see (6.1)) under an equality constraint on $t \in \Delta_N$ equivalent to the periodicity of $\mathcal{U}_N(t)$.

Intuitively, one expects that the results in chapter 5 imply that $\mathcal{F}_N \in C^2(\Delta_N)$. Indeed, only little technical effort is required to prove this. First of all, we recall the definitions in Chapters 5 and 6, but include the dependence of the perturbed functions from r in the notation to avoid confusion. Let $t_0, \ldots, t_N \in [-1, 1]$ with

$$-1 = t_0 < t_1 < \cdots < t_{N-1} < t_N = 1, \tag{8.3}$$

and for $r = (r_1, \ldots, r_{N-1}) \in \mathbb{R}^{N-1}$ we choose $\delta_0 = \delta_0(t, r) > 0$ such that for every $\delta \in (-\delta_0, \delta_0)$, the shifted corners

$$t_i^{(r,\delta)} = t_i + \delta r_i, \quad i = 0, \ldots, N, \tag{8.4}$$

where $r_0 = r_N = 0$ is set as before, satisfy

$$-1 = t_0^{(r,\delta)} < t_1^{(r,\delta)} < \cdots < t_{N-1}^{(r,\delta)} < t_N^{(r,\delta)} = 1. \tag{8.5}$$

If $u \in \mathcal{S}(-1, 1)$ with

$$u'(x) = (-1)^{i-1} \quad \text{for} \quad x \in (t_{i-1}, t_i)$$

for $i = 1, \ldots, N$, we define the perturbed function $u_{r,\delta}$ as the unique element in $\mathcal{S}(-1, 1)$ satisfying

$$
\begin{aligned}
u_{r,\delta}(-1) &= u(-1) \\
u_{r,\delta}'(x) &= (-1)^{i-1} \quad \text{for} \quad x \in (t_{i-1}^{(r,\delta)}, t_i^{(r,\delta)}), \quad i = 1, \ldots, N
\end{aligned}
\tag{8.6}
$$

for every $\delta \in (-\delta_0, \delta_0)$. The function $F_{u,r} : (-\delta_0, \delta_0) \to \mathbb{R}$ is defined by

$$F_{u,r}(\delta) = E(u_{r,\delta}) \quad \text{for every} \quad \delta \in (-\delta_0, \delta_0), \tag{8.7}$$

and as shown in Chapter 5, $F_{u,r} \in C^2(-\delta_0, \delta_0)$ with

$$F_{u,r}'(\delta) = 4 \sum_{i=1}^{N} \alpha_i h_{u,r}^{(i)}(\delta), \tag{8.8}$$

where $\alpha_i \in \mathbb{R}$, $h_{u,r}^{(i)} : (-\delta_0, \delta_0) \to \mathbb{R}$ are given by

$$\alpha_i = \sum_{j=1}^{i-1} (-1)^{j-1} r_j, \quad h_{u,r}^{(i)}(\delta) = \left(\chi_{(t_{i-1}^{(r,\delta)}, t_i^{(r,\delta)})}, u_{r,\delta} \right)_g \tag{8.9}$$

for $i = 1, \ldots, N$, $h_{u,r}^{(i)}$ satisfying (see Theorem 5.4)

$$
\begin{aligned}
(h_{u,r}^{(i)})'(\delta) &= 2 \sum_{k=1}^{N} \alpha_k \left(\chi_{(t_{k-1}^{(r,\delta)}, t_k^{(r,\delta)})}, \chi_{(t_{i-1}^{(r,\delta)}, t_i^{(r,\delta)})} \right)_g \\
&\quad + 2 r_i \int_{-1}^{1} g(t_i^{(r,\delta)} - y)(u_{r,\delta}(t_i^{(r,\delta)}) - u_{r,\delta}(y)) dy \\
&\quad - 2 r_{i-1} \int_{-1}^{1} g(t_{i-1}^{(r,\delta)} - y)(u_{r,\delta}(t_{i-1}^{(r,\delta)}) - u_{r,\delta}(y)) dy
\end{aligned}
\tag{8.10}
$$

Lemma 8.1 *Let $N \in \mathbb{N}$. For given $(t_1, \ldots, t_{N-1}) \in \Delta_N$, $r = (r_1, \ldots, r_{N-1}) \in \mathbb{R}^{N-1}$, let $\delta_0 > 0$ such that $t + \delta r \in \Delta_N$ for every $\delta \in (-\delta_0, \delta_0)$. Then*

$$\mathcal{F}_N(t + \delta r) = F_{\mathcal{U}_N(t), r}(\delta)$$

for every $\delta \in (-\delta_0, \delta_0)$, where $F_{\mathcal{U}_N(t), r}$ is defined as in (5.6).

Proof: We choose δ_0 such that $t + \delta r \in \Delta_N$ for every $\delta \in (-\delta_0, \delta_0)$ which is possible since Δ_N is open. Recalling (8.4), for every $\delta \in (-\delta_0, \delta_0)$ we have

$$t + \delta r = (t_1 + \delta r_1, \ldots, t_{N-1} + \delta r_{N-1}) = (t_1^{(r,\delta)}, \ldots, t_{N-1}^{(r,\delta)}),$$

and the monotonicity (8.5) is satisfied since $t + \delta r \in \Delta_N$. Due to (8.7), we have $F_{\mathcal{U}_N(t), r}(\delta) = E((\mathcal{U}_N(t))_{r,\delta})$, and by (8.6) and (8.1), $(\mathcal{U}_N(t))_{r,\delta} \in \mathcal{S}(-1, 1)$ satisfies

$$((\mathcal{U}_N(t))_{r,\delta})'(x) = (-1)^{i-1} \quad \text{for every} \quad x \in (t_{i-1}^{(r,\delta)}, t_i^{(r,\delta)})$$

for $i = 1, \ldots, N$, where $t_0^{(r,\delta)} = -1$, $t_N^{(r,\delta)} = 1$ is set. Thus,

$$((\mathcal{U}_N(t))_{r,\delta})' = (\mathcal{U}_N(t_1^{(r,\delta)}, \ldots, t_{N-1}^{(r,\delta)}))'$$

due to the definition of $\mathcal{U}_N$, and since E is invariant under the addition of constants, we conclude

$$\mathcal{F}_N(t + \delta r) = E(\mathcal{U}_N(t + \delta r)) = E(\mathcal{U}_N(t_1^{(r,\delta)}, \ldots, t_{N-1}^{(r,\delta)})) = E((\mathcal{U}_N(t))_{r,\delta}) = F_{\mathcal{U}_N(t), r}(\delta)$$

for every $\delta \in (-\delta_0, \delta_0)$. $\qquad\qquad\qquad\qquad\qquad\qquad\qquad\qquad\qquad\qquad\qquad\qquad\qquad\qquad\square$

Theorem 8.2 *Let $N \in \mathbb{N}$, $\Delta_N \subset \mathbb{R}^{N-1}$, $\mathcal{F}_N : \Delta_N \to \mathbb{R}$ defined as above. Then $\mathcal{F}_N \in C^2(\Delta_N)$.*

Proof: Let $v \in \mathbb{R}^{N-1}$ arbitrary, $t = (t_1, \ldots, t_{N-1}) \in \Delta_N$. We compute the directional derivative of $\mathcal{F}_N$ with respect to v. By Lemma 8.1, we get

$$\frac{1}{\delta}[\mathcal{F}_N(t + \delta v) - \mathcal{F}_N(t)] = \frac{1}{\delta}[F_{\mathcal{U}_N(t), v}(\delta) - F_{\mathcal{U}_N(t), v}(0)],$$

and using Theorem 5.2, $\delta \to 0$ yields

$$D_v \mathcal{F}_N(t) = F'_{\mathcal{U}_N(t), v}(0).$$

For arbitrary $w \in \mathbb{R}^{N-1}$, we obtain, using (8.8),

$$\frac{1}{\delta}[D_v \mathcal{F}_N(t + \delta w) - D_v \mathcal{F}_N(t)] = \frac{1}{\delta}[F'_{\mathcal{U}_N(t + \delta w), v}(0) - F'_{\mathcal{U}_N(t), v}(0)] =$$

$$= \frac{4}{\delta} \sum_{i=1}^{N} \sum_{j=1}^{i-1} (-1)^{j-1} v_j [h^{(i)}_{\mathcal{U}_N(t+\delta w),v}(0) - h^{(i)}_{\mathcal{U}_N(t),v}(0)].$$

In order to examine this term further, we derive from (8.9) and (8.4)

$$h^{(i)}_{\mathcal{U}_N(t),w}(\delta) = \left(\chi_{(t_{i-1}+\delta w_{i-1}, t_i+\delta w_i)}, (\mathcal{U}_N(t))_{w,\delta} \right)_g,$$

where $(\mathcal{U}_N(t))_{w,\delta}$ by (8.6) and (8.1) satisfies

$$((\mathcal{U}_N(t))_{w,\delta})'(x) = (-1)^{i-1} \quad \text{for} \quad x \in (t_{i-1} + \delta w_{i-1}, t_i + \delta w_i)$$

for $i = 1, \ldots, N$, and $w_0 = w_N = 0$ is set. Thus, $((\mathcal{U}_N(t))_{w,\delta})' = (\mathcal{U}_N(t + \delta w))'$ by (8.1), and since $(u, v) \mapsto (u, v)_g$ is invariant under the addition of constants in either argument, we deduce

$$h^{(i)}_{\mathcal{U}_N(t),w}(\delta) = \left(\chi_{(t_{i-1}+\delta w_{i-1}, t_i+\delta w_i)}, (\mathcal{U}_N(t + \delta w)) \right)_g = h^{(i)}_{\mathcal{U}_N(t+\delta w),v}(0)$$

so that

$$h^{(i)}_{\mathcal{U}_N(t+\delta w),v}(0) = \left(\chi_{(t_{i-1}+\delta w_{i-1}, t_i+\delta w_i)}, \mathcal{U}_N(t + \delta w) \right)_g = h^{(i)}_{\mathcal{U}_N(t),w}(\delta)$$
$$h^{(i)}_{\mathcal{U}_N(t),v}(0) = \left(\chi_{(t_{i-1}, t_i)}, \mathcal{U}_N(t) \right)_g = h^{(i)}_{\mathcal{U}_N(t),w}(0),$$

which implies

$$\frac{1}{\delta} [D_v \mathcal{F}_N(t + \delta w) - D_v \mathcal{F}_N(t)] = \frac{4}{\delta} \sum_{i=1}^{N} \sum_{j=1}^{i-1} (-1)^{j-1} v_j [h^{(i)}_{\mathcal{U}_N(t),w}(\delta) - h^{(i)}_{\mathcal{U}_N(t),w}(0)].$$

Using Theorem 5.4, $\delta \to 0$ yields

$$D_w D_v \mathcal{F}_N(t) = 4 \sum_{i=1}^{N} \sum_{j=1}^{i-1} (-1)^{j-1} v_j (h^{(i)}_{\mathcal{U}_N(t),w})'(0),$$

where by (8.10)

$$(h^{(i)}_{\mathcal{U}_N(t),w})'(0) = 2 \sum_{k=1}^{N} \sum_{l=1}^{k-1} (-1)^{l-1} w_l \left(\chi_{(t_{k-1}, t_k)}, \chi_{(t_{i-1}, t_i)} \right)_g$$
$$+ 2 w_i \int_{-1}^{1} g(t_i - y) [(\mathcal{U}_N(t))(t_i) - (\mathcal{U}_N(t))(y)] dy$$
$$- 2 w_{i-1} \int_{-1}^{1} g(t_{i-1} - y) [(\mathcal{U}_N(t))(t_{i-1}) - (\mathcal{U}_N(t))(y)] dy. \qquad (8.11)$$

In order to show the continuity of $D_w D_v \mathcal{F}_N$, we use Lebesgue's convergence theorem. Let $s, t \in \Delta_N$, $s = (s_1, \ldots, s_{N-1})$, $t = (t_1, \ldots, t_{N-1})$. Set $s_0 = t_0 = -1$, $s_N = t_N = 1$, then for $i = 0, \ldots, N$ we have

$$(\mathcal{U}_N(t))(t_i) - (\mathcal{U}_N(s))(s_i) =$$

$$= \int_{-1}^{t_i} (\mathcal{U}_N(t))'(\xi)d\xi - \int_{-1}^{s_i} (\mathcal{U}_N(s))'(\xi)d\xi + (\mathcal{U}_N(t))(-1) - (\mathcal{U}_N(s))(-1)$$

$$= \sum_{j=1}^{i} \left[\int_{t_{j-1}}^{t_j} (-1)^{j-1} d\xi - \int_{s_{j-1}}^{s_j} (-1)^{j-1} d\xi \right] + (\mathcal{U}_N(t))(-1) - (\mathcal{U}_N(s))(-1)$$

$$= \sum_{j=1}^{i} (-1)^{j-1} (t_j - s_j + s_{j-1} - t_{j-1}) + (\mathcal{U}_N(t))(-1) - (\mathcal{U}_N(s))(-1)$$

$$= 2 \sum_{j=1}^{i-1} (-1)^{j-1} (t_j - s_j) + (-1)^{i-1}(t_i - s_i) + (\mathcal{U}_N(t))(-1) - (\mathcal{U}_N(s))(-1).$$

On the other hand,

$$(\mathcal{U}_N(s))(y) - (\mathcal{U}_N(t))(y) = \int_{-1}^{y} ((\mathcal{U}_N(s))'(\xi) - (\mathcal{U}_N(t))'(\xi))d\xi$$
$$+ (\mathcal{U}_N(s))(-1) - (\mathcal{U}_N(t))(-1)$$

for $y \in [-1, 1]$, so that

$$(\mathcal{U}_N(t))(t_i) - (\mathcal{U}_N(t))(y) = (\mathcal{U}_N(s))(s_i) - (\mathcal{U}_N(s))(y) + \int_{-1}^{y} ((\mathcal{U}_N(s))'(\xi) - (\mathcal{U}_N(t))'(\xi))d\xi$$

$$+ 2 \sum_{j=1}^{i-1} (-1)^{j-1}(t_j - s_j) + (-1)^{i-1}(t_i - s_i), \qquad (8.12)$$

where

$$\left| 2 \sum_{j=1}^{i-1} (-1)^{j-1}(t_j - s_j) + (-1)^{i-1}(t_i - s_i) \right| \le 2 \sum_{j=1}^{i} |t_j - s_j| \to 0$$

for $i = 0, \ldots, N$ if $s \to t$. For fixed $t \in \Delta_N$, we choose $\gamma > 0$ so small that for every $s \in \Delta_N$ with $\|t - s\| \le \gamma$, the condition $s_{i-1} \vee t_{i-1} < s_i \wedge t_i$ holds for $i = 1, \ldots, N$. Then for every $y \in (-1, 1)$, $s \in \Delta_N$ with $\|t - s\| \le \gamma$, $|(\mathcal{U}_N(s))'(y) - (\mathcal{U}_N(t))'(y)| = 0$ holds for $y \in (s_{i-1}, s_i) \cap (t_{i-1}, t_i) = (s_{i-1} \vee t_{i-1}, s_i \wedge t_i)$, $i = 1, \ldots, N$ due to (8.1), and we get

$$\left| \int_{-1}^{y} ((\mathcal{U}_N(s))'(\xi) - (\mathcal{U}_N(t))'(\xi))d\xi \right| \le \int_{-1}^{1} |(\mathcal{U}_N(s))'(\xi) - (\mathcal{U}_N(t))'(\xi)|d\xi$$

$$= 2 \sum_{i=1}^{N-1} \int_{s_i \wedge t_i}^{s_i \vee t_i} d\xi \le 2 \sum_{i=1}^{N-1} |s_i - t_i| \to 0$$

for $s \to t$. Combining this with (8.12), we conclude that for every $y \in [-1, 1]$,

$$(\mathcal{U}_N(s))(s_i) - (\mathcal{U}_N(s))(y) \to (\mathcal{U}_N(t))(t_i) - (\mathcal{U}_N(t))(y) \quad \text{for} \quad s \to t.$$

Since g is continuous, we obtain

$$g(s_i - y)[(\mathcal{U}_N(s))(s_i) - (\mathcal{U}_N(t))(y)] \to g(t_i - y)[(\mathcal{U}_N(t))(t_i) - (\mathcal{U}_N(t))(y)] \quad \text{for} \quad s \to t$$

pointwise. We recall that g is bounded, and since $\int_{-1}^{1}(\mathcal{U}_N(t))(x)dx = 0$, we find $x_0 \in [-1, 1]$ with $(\mathcal{U}_N(t))(x_0) = 0$, so that for every $t \in \Delta_N$,

$$|(\mathcal{U}_N(t))(x)| = \left| \int_{x_0}^{x} (\mathcal{U}_N(t))'(\xi)d\xi \right| \leq |x - x_0| \leq 1$$

for every $x \in [-1, 1]$, which implies the boundedness of the integrands. Thus, the second and third line in (8.11) are continuous in t due to Lebesgue's convergence theorem, as is the first one, which is easy to see, since $\chi_{(s_{i-1}, s_i)} \to \chi_{(t_{i-1}, t_i)}$ pointwise for $s \to t$.

We have shown that for arbitrary $v, w \in \mathbb{R}^{N-1}$, $D_w D_v \mathcal{F}_N(t)$ exists for every $t \in \Delta_N$, and $D_w D_v \mathcal{F}_N : \Delta_N \to \mathbb{R}$ is continuous, which implies $\mathcal{F}_N \in C^2(\Delta_N)$. $\qquad\square$

8.2 Local Minimality of the Nonlocal Energy

Intuitively, one believes that if $(\bar{t}_1, \ldots, \bar{t}_{N-1}) \in \Delta_N$ given by (6.1) is a local minimizer of $\mathcal{F}_N : \Delta_N \to \mathbb{R}$ under the constraint that $\mathcal{U}_N(t)$ is periodic, then the associated sawtooth function $\bar{u}_N \in \mathcal{S}_{per}^0(-1, 1)$ (cf. (6.3), (6.2)) is an H^1-local minimizer of E in $\mathcal{S}_{per}^0(-1, 1)$ in sense of Corollary 4.9, i.e. in the set of all $u \in \mathcal{S}_{per}^0(-1, 1)$ with N corners.

First of all, we will draw the connection between 0 as a local minimum point of any "r-perturbation" $\bar{F}_r$ of $\bar{u}_N$ with r satisfying (6.8) and the "corner vector" $(\bar{t}_1, \ldots, \bar{t}_{N-1})$ as a local minimum point of $\mathcal{F}_N$ under the constraint of periodicity. Latter is expressed in "Δ_N-calculus" as follows:

Lemma 8.3 *For $N \in \mathbb{N}$ even, let $t = (t_1, \ldots, t_{N-1}) \in \Delta_N$. Then*

$$\mathcal{U}_N(t) \in \mathcal{S}_{per}^0(-1, 1) \Leftrightarrow \sum_{i=1}^{N-1}(-1)^i t_i = 0. \tag{8.13}$$

Proof: Using (8.1) and the fact that N is even, we obtain

$$(\mathcal{U}_N(t))(1) - (\mathcal{U}_N(t))(-1) = \int_{-1}^{1}(\mathcal{U}_N(t))'(\xi)d\xi = \sum_{i=1}^{N}\int_{t_{i-1}}^{t_i}(-1)^{i-1}d\xi$$

$$= \sum_{i=1}^{N}(-1)^{i-1}(t_i - t_{i-1}) = \sum_{i=1}^{N}(-1)^{i-1}t_i + \sum_{i=0}^{N-1}(-1)^{i-1}t_i$$

$$= -t_0 - t_N + 2\sum_{i=1}^{N-1}(-1)^{i-1}t_i = 2\sum_{i=1}^{N-1}(-1)^{i-1}t_i$$

since $t_0 = -1$, $t_N = 1$. (8.13) now immediately follows. $\qquad\square$

Theorem 8.4 *Let $N \in \mathbb{N}$ be an even number, $A_N \in \mathbb{R}^{N \times N}$ given in Theorem 6.4, and let $\bar{t}_i$, $i = 0, \ldots, N$ as given by (6.1). Assume that for every $(r_1, \ldots, r_{N-1}) \in \mathbb{R}^{N-1}$ satisfying the periodicity condition (6.8), we have*

$$\bar{F}_r'(0) = 0, \quad \bar{F}_r''(0) > 0.$$

Then

$$\mathcal{F}_N : \Delta_N \to \mathbb{R}$$

attains a strict local minimum at $\bar{t} = (\bar{t}_1, \ldots, \bar{t}_{N-1}) \in \Delta_N$ under the condition

$$\mathcal{H}_N(t) = 0,$$

where $\mathcal{H}_N : \Delta_N \to \mathbb{R}$ is defined by

$$\mathcal{H}_N(t) = \sum_{i=1}^{N-1} (-1)^i t_i \quad \text{for every} \quad t \in \Delta_N.$$

Proof: Using Ljusternik's sufficient condition for local minima under equality constraints (see [48], Theorem 43.D), we just need to show that we find $\lambda \in \mathbb{R}$ such that

$$D\mathcal{F}_N(\bar{t}) = \lambda D\mathcal{H}_N(\bar{t}) = (-\lambda, \lambda, -\lambda, \ldots, -\lambda) \tag{8.14}$$

and $D^2(\mathcal{F}_N - \lambda \mathcal{H}_N)(\bar{t}) = D^2 \mathcal{F}_N(\bar{t})$ (which denotes the Hessian of $\mathcal{F}_N$) is positive definite on the set

$$\{r \in \mathbb{R}^{N-1} \mid r \perp D\mathcal{H}_N(\bar{t})\} = \{r \in \mathbb{R}^{N-1} \mid r \perp (-1, 1, -1, \ldots, -1)\},$$

or, equivalently,

$$r^t[D^2 \mathcal{F}_N(\bar{t})]r > 0 \quad \text{for every } r = (r_1, \ldots, r_{N-1}) \in \mathbb{R}^{N-1} \text{ with } \sum_{i=1}^{N-1} (-1)^i r_i = 0. \tag{8.15}$$

Recalling the definition of $\bar{u}_N$ (cf. (6.3), (6.2)), we easily deduce from (8.1) that $(\mathcal{U}_N(\bar{t}))' = \bar{u}_N'$, which by (8.6) implies $((\mathcal{U}_N(\bar{t}))_{r,\delta})' = ((\bar{u}_N)_{r,\delta})'$. Using Lemma 8.1 and the invariance of E under the addition of constants, we obtain

$$\mathcal{F}_N(\bar{t} + \delta r) = F_{\mathcal{U}_N(\bar{t}),r}(\delta) = E((\mathcal{U}_N(\bar{t}))_{r,\delta}) = E((\bar{u}_N)_{r,\delta}) = F_{\bar{u}_N,r}(\delta) = \bar{F}_r(\delta)$$

which implies

$$\bar{F}_r'(0) = D\mathcal{F}_N(\bar{t})r, \quad \bar{F}_r''(0) = r^t[D^2 \mathcal{F}_N(\bar{t})]r.$$

Since $\bar{F}_r'(0) = 0$ for every $r \in \mathbb{R}^{N-1}$ satisfying (6.8), which in particular is valid for $r_i = e_i + e_{i+1}$, $i = 1, \ldots, N-2$ (where $e_i = (0, \ldots, 0, 1, 0, \ldots, 0)$ denotes the i-th unit vector in $\mathbb{R}^{N-1}$), we obtain

$$0 = \bar{F}_{r_i}'(0) = D\mathcal{F}_N(\bar{t})r_i = \frac{\partial \mathcal{F}_N}{\partial t_i}(\bar{t}) + \frac{\partial \mathcal{F}_N}{\partial t_{i+1}}(\bar{t})$$

for $i = 1, \ldots, N - 2$, which clearly implies (8.14). For every $r \in \mathbb{R}^{N-1}$ satisfying (6.8), we know that

$$r^t [D^2 \mathcal{F}_N(\bar{t})] r = \bar{F}_r''(0) > 0,$$

which is equivalent to (8.15). $\square$

Theorem 8.5 *Let $N \in \mathbb{N}$ be an even number, $\mathcal{F}_N$, $\mathcal{H}_N : \Delta_N \to \mathbb{R}$ defined as before, and $\bar{t}_i$, $i = 0, \ldots, N$ given by (6.1). Assume that $\mathcal{F}_N$ has a strict local minimum at $(\bar{t}_1, \ldots, \bar{t}_{N-1}) \in \Delta_N$ under the condition*

$$\mathcal{H}_N(t) = 0.$$

Then there exists $\delta > 0$ such that for every $u \in \mathcal{S}_{per}^0(-1, 1)$ with $\#(Su' \cap [-1, 1)) = N$, $u \notin \mathcal{O}(\bar{u}_N)$ and $\|u - \bar{u}_N\|_{H^1(-1,1)} < \delta$, we have

$$E(u) > E(\bar{u}_N),$$

where $\bar{u}_N \in \mathcal{S}_{per}^0(-1, 1)$ is the function defined by (6.3), (6.2). In other words, $\bar{u}_N$ is an isolated H^1-local minimizer of E "up to translation" in the set of all $u \in \mathcal{S}_{per}^0(-1, 1)$ with N corners on $[-1, 1)$.

Proof: Assume that this is not the case. Then we will find a sequence $(u_n)_{n \in \mathbb{N}} \subset \mathcal{S}_{per}^0(-1, 1)$ with $\#(Su_n' \cap [-1, 1)) = N$ and $u_n \notin \mathcal{O}(\bar{u}_N)$ for every $n \in \mathbb{N}$ such that

$$\lim_{n \to \infty} \|u_n - \bar{u}_N\|_{H^1(-1,1)} = 0$$

and

$$E(u_n) \leq E(\bar{u}_N) \quad \text{for every} \quad n \in \mathbb{N}.$$

In particular, we have

$$\lim_{n \to \infty} \|u_n' - \bar{u}_N'\|_{L^2(-1,1)} = 0.$$

For every $n \in \mathbb{N}$, let

$$-1 \leq t_1^{(n)} < \cdots < t_N^{(n)} < 1$$

be the N corners of u_n in $[-1, 1)$, so that for every $n \in \mathbb{N}$, we have

$$u_n'(x) = \sigma_n(-1)^i \quad \text{for} \quad x \in (t_{i-1}^{(n)}, t_i^{(n)}), \quad i = 1, \ldots, N+1, \tag{8.16}$$

where $t_0^{(n)} = -1$, $t_{N+1}^{(n)} = 1$ is set, and $\sigma_n \in \{\pm 1\}$ for every $n \in \mathbb{N}$. By choosing a subsequence, we may assume that

$$\lim_{n \to \infty} t_i^{(n)} = t_i \in [-1, 1] \quad \text{for} \quad i = 0, \ldots, N+1, \tag{8.17}$$

where

$$-1 = t_0 \leq t_1 \leq \cdots \leq t_N \leq t_{N+1} = 1.$$

We may also assume that $\sigma_n = 1$ for every $n \in \mathbb{N}$ or $\sigma_n = -1$ for every $n \in \mathbb{N}$. In particular, $\lim_{n \to \infty} \sigma_n = \sigma \in \{\pm 1\}$. Clearly, (8.17) and (8.16) now imply

$$\lim_{n \to \infty} u'_n(x) = v(x) \quad \text{for a.e.} \quad x \in [-1, 1],$$

where

$$v(x) = \sigma(-1)^i \quad \text{for} \quad x \in (t_{i-1}, t_i), \quad i = 1, \ldots, N+1.$$

In particular, $\lim_{n \to \infty} \|u'_n - v\|_{L^2(-1,1)} = 0$, so that $v = \bar{u}'_N$. Since

$$\bar{u}'_N(x) = (-1)^{i-1} \quad \text{for} \quad x \in (\bar{t}_{i-1}, \bar{t}_i), \quad i = 1, \ldots, N$$

(cf. (6.3), (6.2)), we may assume that

$$\sigma = -1, \quad t_i = \bar{t}_i \quad \text{for} \quad i = 0, \ldots, N, \quad t_{N+1} = t_N = 1.$$

This implies that $\sigma_n = -1$ for every $n \in \mathbb{N}$, so that (8.16) becomes

$$u'_n(x) = (-1)^{i-1} \quad \text{for} \quad x \in (t_{i-1}^{(n)}, t_i^{(n)}), \quad i = 1, \ldots, N,$$

and from (8.17) we deduce

$$\lim_{n \to \infty} t_i^{(n)} = \bar{t}_i \quad \text{for} \quad i = 1, \ldots, N. \tag{8.18}$$

We replace u_n (i.e. its periodic extension to $\mathbb{R}$) by

$$\tilde{u}_n = T_{\tau_n} u_n = u_n(\cdot - \tau_n), \quad \text{where} \quad \tau_n = 1 - t_N^{(n)} > 0$$

(cf. Definition 4.3). Obviously, $(\tilde{u}_n)_{n \in \mathbb{N}} \subset \mathcal{S}_{per}^0(-1, 1)$ with $\tilde{u}_n \notin \mathcal{O}(\bar{u}_N)$ for every $n \in \mathbb{N}$. Since $\tilde{u}'_n = u'_n(\cdot - \tau_n)$, for every n we obtain

$$\tilde{u}'_n(x) = (-1)^{i-1} \quad \text{for} \quad x \in (\tilde{t}_{i-1}^{(n)}, \tilde{t}_i^{(n)}), \quad i = 2, \ldots, N$$

with $\tilde{t}_i^{(n)} = t_i^{(n)} + \tau_n = t_i^{(n)} + 1 - t_N^{(n)}$ which implies $\tilde{t}_N^{(n)} = 1$. Since the 2-periodic extension of u_n satisfies $u'_n(x) = 1$ on $(t_N^{(n)} - 2, t_1^{(n)})$, we deduce

$$\tilde{u}'_n(x) = 1 \quad \text{for} \quad x \in (t_N^{(n)} - 2 + \tau_n, t_1^{(n)} + \tau_n) = (-1, \tilde{t}_1^{(n)}).$$

Combining these properties, for every $n \in \mathbb{N}$ we obtain

$$\tilde{u}'_n(x) = (-1)^{i-1} \quad \text{for} \quad x \in (\tilde{t}_{i-1}^{(n)}, \tilde{t}_i^{(n)}), \quad i = 1, \ldots, N \tag{8.19}$$

with

$$-1 = \tilde{t}_0^{(n)} < \tilde{t}_1^{(n)} < \cdots < \tilde{t}_{N-1}^{(n)} < \tilde{t}_N^{(n)} = 1.$$

Thus, $(\tilde{t}_1^{(n)}, \ldots, \tilde{t}_{N-1}^{(n)}) \in \Delta_N$ with $\tilde{u}_n = \mathcal{U}_N(\tilde{t}_1^{(n)}, \ldots, \tilde{t}_{N-1}^{(n)})$ (cf. (8.1)), and since $\tilde{u}_n$ lie in $\mathcal{S}_{per}^0(-1, 1)$, Lemma 8.3 implies that

$$\mathcal{H}_N(\tilde{t}_1^{(n)}, \ldots, \tilde{t}_{N-1}^{(n)}) = \sum_{i=1}^{N-1} (-1)^i \tilde{t}_i^{(n)} = 0$$

for every $n \in \mathbb{N}$. From the definitions (8.2) and (8.1), we conclude, using the invariance of E under translations,

$$\begin{aligned} \mathcal{F}_N(\tilde{t}_1^{(n)}, \ldots, \tilde{t}_N^{(n)}) &= E(\mathcal{U}_N(\tilde{t}_1^{(n)}, \ldots, \tilde{t}_N^{(n)})) = E(\tilde{u}_n) = E(T_{\tau_n} u_n) \\ &= E(u_n) \leq E(\bar{u}_N) = E(\mathcal{U}_N(\bar{t}_1, \ldots, \bar{t}_N)) = \mathcal{F}_N(\bar{t}_1, \ldots, \bar{t}_N). \end{aligned}$$

Furthermore,

$$\lim_{n \to \infty} \tilde{t}_i^{(n)} = \lim_{n \to \infty} (t_i^{(n)} + 1 - t_N^{(n)}) = \bar{t}_i + 1 - \bar{t}_N = \bar{t}_i \quad \text{for} \quad i = 1, \ldots, N$$

due to (8.18) and (6.1), and for every $n \in \mathbb{N}$ we have

$$(\tilde{t}_1^{(n)}, \ldots, \tilde{t}_{N-1}^{(n)}) \neq (\bar{t}_1, \ldots, \bar{t}_{N-1}),$$

since otherwise (8.19), (6.3) would imply $\tilde{u}_n' = \bar{u}_N'$, which due to $\int \bar{u}_N = \int \tilde{u}_n = 0$ yields $\tilde{u}_n = \bar{u}_N \in \mathcal{O}(\bar{u}_N)$, a contradiction.
Combining all these results, we obtain a sequence

$$((\tilde{t}_1^{(n)}, \ldots, \tilde{t}_{N-1}^{(n)}))_{n \in \mathbb{N}} \subset \Delta_N, \quad (\tilde{t}_1^{(n)}, \ldots, \tilde{t}_{N-1}^{(n)}) \neq (\bar{t}_1, \ldots, \bar{t}_{N-1}) \quad \text{for every} \quad n \in \mathbb{N}$$

with

$$\lim_{n \to \infty} (\tilde{t}_1^{(n)}, \ldots, \tilde{t}_{N-1}^{(n)}) = (\bar{t}_1, \ldots, \bar{t}_{N-1})$$

and

$$\mathcal{H}_N(\tilde{t}_1^{(n)}, \ldots, \tilde{t}_{N-1}^{(n)}) = 0, \quad \mathcal{F}_N(\tilde{t}_1^{(n)}, \ldots, \tilde{t}_{N-1}^{(n)}) \leq \mathcal{F}_N(\bar{t}_1, \ldots \bar{t}_{N-1}) \quad \text{for every} \quad n \in \mathbb{N}.$$

This contradicts the assumption that $\mathcal{F}_N$ has a strict local minimum at $(\bar{t}_1, \ldots, \bar{t}_{N-1})$ in Δ_N under the condition $\mathcal{H}_N(t) = 0$. $\qquad \square$

8.3 Connection to the Eigenvalues of A_N

We can now easily show how positivity of eigenvalues of A_N implies that the equidistant sawtooth function $\bar{u}_N$ with N corners in $[-1, 1)$ is indeed an H^1-local minimizer of the Γ-limit I, which by Corollary 4.9 delivers local minimizers of I^ε converging to $\bar{u}_N$.

Corollary 8.6 *For an even number $N \in \mathbb{N}$, let A_N as given in Theorem 6.4 with eigenvalues $\lambda_t^{(N)}$, $t = 0, \ldots, \frac{N}{2}$, given by (7.1). Let $\bar{u}_N \in \mathcal{S}_{per}^0(-1, 1)$ as defined by (6.3), (6.2). Assume that*

$$\lambda_t^{(N)} > 0 \quad \text{for} \quad t = 1, \ldots, \frac{N}{2} - 1.$$

Then the following holds

(i) *There exists $\delta > 0$ such that for every $u \in \mathcal{S}_{per}^0(-1, 1)$ with $\|u - \bar{u}_N\|_{H^1(-1,1)} < \delta$, $u \notin \mathcal{O}(\bar{u}_N)$ and $\#(Su' \cap [-1, 1)) = N$, we have*

$$E(u) > E(\bar{u}_N).$$

(ii) *There exists $\delta > 0$ such that for every $u \in \mathcal{S}^0_{per}(-1,1)$ with $\|u - \bar{u}_N\|_{H^1(-1,1)} < \delta$ and $u \notin \mathcal{O}(\bar{u}_N)$, we have*

$$I(u) > I(\bar{u}_N).$$

(iii) *There exist $\delta > 0$, $\varepsilon_0 > 0$ such that for every $\varepsilon \in (0, \varepsilon_0)$ we find $u_\varepsilon \in H^{2,0}_{per}(-1,1)$ with $\|u_\varepsilon - \bar{u}_N\|_{H^1(-1,1)} < \delta$ and*

$$I^\varepsilon(u_\varepsilon) \leq I^\varepsilon(u) \quad \text{for every} \quad u \in H^{2,0}_{per}(-1,1) \quad \text{with} \quad \|u - \bar{u}_N\|_{H^1(-1,1)} < \delta.$$

Furthermore,

$$\lim_{\varepsilon \to 0} \|u_\varepsilon - \bar{u}_N\|_{H^1(-1,1)} = 0.$$

In particular, for every $\varepsilon \in (0, \varepsilon_0)$, u_ε is a local $H^1(-1,1)$-minimizer of I^ε in the space $H^{2,0}_{per}(-1,1)$.

Proof: Apply Theorems 7.4, 8.4, 8.5 to show (i). Then (ii) immediately follows from Theorem 4.8. Application of Corollary 4.9 to (i) finally yields (iii). □

This result justifies our aim to show that for large N, the eigenvalues $\lambda^{(N)}_1, \ldots, \lambda^{(N)}_{\frac{N}{2}-1}$ are positive which will be done in the following two chapters. As a first step, we will show that we only have to consider eigenvalues $\lambda^{(N)}_t$ with $\frac{t}{N} \approx \frac{1}{2}$. The second step is to estimate these eigenvalues.

Chapter 9

Asymptotic Behavior of Eigenvalues

9.1 Basic Estimates

Since Corollary 8.6 shows how positivity of the eigenvalues $\lambda_t^{(N)}$, $t = 1, \ldots, \frac{N}{2} - 1$, leads to local minimizers $u_\varepsilon \in H_{per}^{2,0}(-1,1)$ of the functionals I^ε for small $\varepsilon > 0$, we will deal with these eigenvalues in this and the next chapter. Recalling (7.1), for every even $N \in \mathbb{N}$ these are given by

$$\lambda_t^{(N)} = a_0^{(N)} + (-1)^t a_{\frac{N}{2}}^{(N)} + 2 \sum_{k=1}^{\frac{N}{2}-1} a_k^{(N)} \cos\left[\frac{2\pi kt}{N}\right] \quad \text{for} \quad t = 0, \ldots, \frac{N}{2}, \qquad (9.1)$$

the coefficients $a_k^{(N)}$, $k = 0, \ldots, \frac{N}{2}$, given as in Theorem 6.4, and for every $t \in \left\{0, \ldots, \frac{N}{2}\right\}$, a corresponding eigenvector for $\lambda_t^{(N)}$ due to (7.2) is given by

$$v_t = (v_t^{(0)}, \ldots, v_t^{(N-1)}) \in \mathbb{R}^N,$$
$$v_t^{(k)} = \cos\left[\frac{2\pi kt}{N}\right], \quad k = 0, \ldots, N - 1. \qquad (9.2)$$

We will compare these eigenvalues to those of a different symmetric circulant matrix $\bar{A}_N = (\bar{a}_{|i-j|}^{(N)})_{i,j=1,\ldots,N}$ defined by

$$\bar{A}_N = A_N - D_N = B_N + C_N \in \mathbb{R}^{N \times N}, \qquad (9.3)$$

where $B_N = (b_{|i-j|}^{(N)})_{i,j=1,\ldots,N}$, $C_N = (c_{|i-j|}^{(N)})_{i,j=1,\ldots,N} \in \mathbb{R}^{N \times N}$, $D_N = (d_{|i-j|}^{(N)})_{i,j=1,\ldots,N} \in \mathbb{R}^{N \times N}$ are the symmetric circulant matrices given in Theorem 6.4. Obviously, we have

$$\begin{array}{rcll}
\bar{a}_0^{(N)} & = & b_0^{(N)} + c_0^{(N)} & \\
\bar{a}_1^{(N)} = \bar{a}_{N-1}^{(N)} & = & b_1^{(N)} + c_1^{(N)} & \\
\bar{a}_k^{(N)} & = & c_k^{(N)} & \text{for} \quad k = 2, \ldots, N - 2.
\end{array} \qquad (9.4)$$

Applying Theorem 7.2, we can write down the eigenvalues of $\bar{A}_N$. These are given by

$$\bar{\lambda}_t^{(N)} = \bar{a}_0^{(N)} + (-1)^t \bar{a}_{\frac{N}{2}}^{(N)} + 2 \sum_{k=1}^{\frac{N}{2}-1} \bar{a}_k^{(N)} \cos\left[\frac{2\pi kt}{N}\right] \quad \text{for} \quad t = 0, \ldots, \frac{N}{2} \qquad (9.5)$$

with corresponding eigenvectors

$$v_t = (v_t^{(0)}, \dots, v_t^{(N-1)}) \in \mathbb{R}^N,$$

$$v_t^{(k)} = \cos\left[\frac{2\pi kt}{N}\right], \quad k = 0, \dots, N-1. \tag{9.6}$$

We have the following estimate:

Lemma 9.1 *There exists a constant $C > 0$, only dependent on g, such that for every even $N \in \mathbb{N}$, we have*

$$\left|\lambda_t^{(N)} - \bar{\lambda}_t^{(N)}\right| \leq \frac{C}{N} \tag{9.7}$$

for $t = 0, \dots, \frac{N}{2}$.

Proof: From (9.1), (9.5), (9.4) and the definitions in Theorem 6.4 we conclude

$$
\begin{aligned}
\lambda_t^{(N)} - \bar{\lambda}_t^{(N)} &= (a_0^{(N)} - \bar{a}_0^{(N)}) + (-1)^t(a_{\frac{N}{2}}^{(N)} - \bar{a}_{\frac{N}{2}}^{(N)}) + 2\sum_{k=1}^{\frac{N}{2}-1}(a_k^{(N)} - \bar{a}_k^{(N)})\cos\left[\frac{2\pi kt}{N}\right] \\
&= d_0^{(N)} + 2d_1^{(N)}\cos\left[\frac{2\pi t}{N}\right] = d_0^{(N)}\left[1 - \cos\left[\frac{2\pi t}{N}\right]\right] \\
&= N(-1)^{\frac{N}{2}+1}\int_{-1}^{1} g(y)\bar{u}_N(y)dy\left[1 - \cos\left[\frac{2\pi t}{N}\right]\right]
\end{aligned}
$$

Recalling (6.1), $\int_{\bar{t}_{i-1}}^{\bar{t}_i} \bar{u}_N(y)dy = 0$ for $i = 1, \dots, N$ and $|\bar{u}_N| \leq \frac{1}{N}$ (cf. Lemma 6.1), we obtain

$$
\begin{aligned}
\left|\lambda_t^{(N)} - \bar{\lambda}_t^{(N)}\right| &\leq 2N\left|\int_{-1}^{1} g(y)\bar{u}_N(y)dy\right| = 2N\left|\sum_{i=1}^{N}\int_{\bar{t}_{i-1}}^{\bar{t}_i} g(y)\bar{u}_N(y)dy\right| \\
&= 2N\left|\sum_{i=1}^{N}\int_{\bar{t}_{i-1}}^{\bar{t}_i}(g(y) - g(t_{i-1}))\bar{u}_N(y)dy\right| \leq 2\sum_{i=1}^{N}\int_{\bar{t}_{i-1}}^{\bar{t}_i}|g(y) - g(t_{i-1})|dy.
\end{aligned}
$$

For $y \in (\bar{t}_{i-1}, \bar{t}_i)$, $i = 1, \dots, N$, we observe that due to the Lipschitz continuity of g,

$$|g(y) - g(t_{i-1})| \leq C|y - t_{i-1}| \leq C|t_i - t_{i-1}| = \frac{C}{N},$$

which implies

$$\left|\lambda_t^{(N)} - \bar{\lambda}_t^{(N)}\right| \leq \frac{C}{N}\sum_{i=1}^{N}(\bar{t}_i - \bar{t}_{i-1}) = \frac{C}{N}(\bar{t}_N - \bar{t}_0) \leq \frac{C}{N}.$$

$\square$

Lemma 9.2 *There is a constant $C > 0$, only dependent on g, such that for every even $N \in \mathbb{N}$, we have*

$$|\lambda_t^{(N)}|, \; |\bar{\lambda}_t^{(N)}| \leq C \tag{9.8}$$

for $t = 0, \ldots, \frac{N}{2}$.

Proof: It is sufficient to show the estimate for $\bar{\lambda}_t^{(N)}$, the other will then follow directly from Lemma 9.1. Fix an even $N \in \mathbb{N}$ and let $\bar{\lambda}_t^{(N)}$, $t \in \{0, \ldots, \frac{N}{2}\}$, be an eigenvalue of $\bar{A}_N$ and $\alpha = (\alpha_1, \ldots, \alpha_N) \in \mathbb{R}^N$ a corresponding normed eigenvector of A_N (i.e. $\|\alpha\| = 1$) and set $\alpha_{N+1} = \alpha_1$. Then, using (9.5), (6.11) and $b_0^{(N)} > 0$ (which follows from $g > 0$),

$$
\begin{aligned}
\bar{\lambda}_t^{(N)} &= \bar{\lambda}_t^{(N)} \alpha^t \alpha = \alpha^t \bar{A}_N \alpha = \alpha^t B_N \alpha + \alpha^t C_N \alpha \\
&= b_0^{(N)} \sum_{i=1}^{N} \alpha_i^2 + 2b_1^{(N)} \sum_{i=1}^{N-1} \alpha_i \alpha_{i+1} + 2b_1^{(N)} \alpha_N \alpha_1 + \sum_{i=1}^{N} \sum_{j=1}^{N} \alpha_i \alpha_j c_{ij}^{(N)} \\
&= b_0^{(N)} \sum_{i=1}^{N} (\alpha_i^2 + \alpha_i \alpha_{i+1}) + \sum_{i=1}^{N} \sum_{j=1}^{N} \alpha_i \alpha_j c_{ij}^{(N)} \\
&\geq b_0^{(N)} \sum_{i=1}^{N} (\alpha_i^2 - |\alpha_i \alpha_{i+1}|) - \sum_{i=1}^{N} \sum_{j=1}^{N} |\alpha_i||\alpha_j||c_{ij}^{(N)}| .
\end{aligned} \tag{9.9}
$$

From the Cauchy-Schwarz inequality, we obtain, using $\alpha_{N+1} = \alpha_1$

$$
\sum_{i=1}^{N} (\alpha_i^2 - |\alpha_i \alpha_{i+1}|) \geq \left(\sum_{i=1}^{N} \alpha_i^2 - \left(\sum_{i=1}^{N} \alpha_i^2 \right)^{\frac{1}{2}} \left(\sum_{i=1}^{N} \alpha_{i+1}^2 \right)^{\frac{1}{2}} \right) = \sum_{i=1}^{N} \alpha_i^2 - \sum_{i=1}^{N} \alpha_i^2 = 0 .
$$

We use (6.13) and the boundedness of g to deduce

$$
|c_{ij}^{(N)}| \leq N \int_{\frac{2}{N}(i-1)}^{\frac{2}{N}i} \int_{\frac{2}{N}(j-1)}^{\frac{2}{N}j} |g(x-y)| dy dx \leq CN \left(\frac{2}{N} \right)^2 \leq \frac{C}{N} \tag{9.10}
$$

for every i, $j = 1, \ldots, N$, and since $b_0^{(N)} > 0$ we conclude, using the Cauchy-Schwarz inequality and $\|\alpha\| = 1$,

$$
\bar{\lambda}_t^{(N)} \geq -\frac{C}{N} \sum_{i=1}^{N} \sum_{j=1}^{N} |\alpha_i||\alpha_j| = -\frac{C}{N} \left(\sum_{i=1}^{N} |\alpha_i| \right)^2 \geq -\frac{C}{N} \left(\sum_{i=1}^{N} |\alpha_i|^2 \right) \left(\sum_{i=1}^{N} 1^2 \right) = -C ,
$$

which shows the lower bound. To show the upper bound, we use (9.9), (9.10), $\alpha_{N+1} = \alpha_1$, $\|\alpha\| = 1$ and the Cauchy-Schwarz inequality and obtain

$$
\begin{aligned}
\bar{\lambda}_t^{(N)} &= b_0^{(N)} \sum_{i=1}^{N} (\alpha_i^2 + \alpha_i \alpha_{i+1}) + \sum_{i=1}^{N} \sum_{j=1}^{N} \alpha_i \alpha_j c_{ij}^{(N)} \\
&\leq b_0^{(N)} \left(\sum_{i=1}^{N} \alpha_i^2 + \left(\sum_{i=1}^{N} \alpha_i^2 \right)^{\frac{1}{2}} \left(\sum_{i=1}^{N} \alpha_{i+1}^2 \right)^{\frac{1}{2}} \right) + \frac{C}{N} \sum_{i=1}^{N} \sum_{j=1}^{N} |\alpha_i||\alpha_j|
\end{aligned}
$$

$$\begin{aligned} &= 2b_0^{(N)} \sum_{i=1}^{N} \alpha_i^2 + \frac{C}{N} \left(\sum_{i=1}^{N} |\alpha_i| \right)^2 \leq 2b_0^{(N)} + \frac{C}{N} \left(\sum_{i=1}^{N} |\alpha_i|^2 \right) \left(\sum_{i=1}^{N} 1^2 \right) \\ &= 2b_0^{(N)} + C \leq C, \end{aligned}$$

and the proof of the Lemma is complete. $\qquad\square$

9.2 Non-Positive Limits of Eigenvalues

In what follows, we will show that a sequence of eigenvalues $(\lambda_{t_n}^{(N_n)})_{n\in\mathbb{N}}$ with $N_n \nearrow \infty$ for $n \to \infty$ and $t_n \in \{1, \ldots, \frac{N_n}{2} - 1\}$ which in the limit $n \to \infty$ are non-positive can only converge to 0, and that $\frac{t_n}{N_n} \approx \frac{1}{2}$ for large n. Due to Lemma 9.1, we may as well consider the eigenvalues of $\bar{A}_{N_n}$. The idea is then to re-write the bilinear form associated with $\bar{A}_{N_n}$ in terms of discrete integrals over simple functions defined by the vector components. In the limit $n \to \infty$, we obtain a bilinear form $\mathcal{B}$ that satisfies a certain positivity condition. We derive an abstract weak convergence property (Theorem 9.4), from which we will deduce that the "C_N-part" of the Fourier sum in (9.5) converges to zero. First of all, we will need the following property:

Lemma 9.3 Let $\mathcal{B} : L^2(0,2) \times L^2(0,2) \to \mathbb{R}$ given by

$$\mathcal{B}(\psi, \varphi) = 2 \int_{-1}^{1} g(y) dy \int_{0}^{2} \psi(x)\varphi(x) dx - 2 \int_{0}^{2} \int_{0}^{2} g(x-y)\psi(x)\varphi(y) dy dx \,. \qquad (9.11)$$

Then $\mathcal{B}$ is a continuous, symmetric bilinear form.

Proof: For given φ, $\psi \in L^2(0,2)$, we can estimate $\mathcal{B}(\psi, \varphi)$ as follows, using the boundedness of g and the Cauchy-Schwarz inequality:

$$\begin{aligned} |\mathcal{B}(\psi, \varphi)| &\leq C|(\psi, \varphi)_{L^2(0,2)}| + C \int_{0}^{2} \int_{0}^{2} |\psi(x)||\varphi(y)| dy dx \\ &\leq C\|\psi\|_{L^2(0,2)}\|\varphi\|_{L^2(0,2)} + C\|\psi\|_{L^1(0,2)}\|\varphi\|_{L^1(0,2)} \leq C\|\psi\|_{L^2(0,2)}\|\varphi\|_{L^2(0,2)} \,. \end{aligned}$$

Thus, $\mathcal{B}(\psi, \varphi)$ is well-defined. The linearity in either argument is trivial, as is the symmetry, and the continuity immediately follows from above estimate. $\qquad\square$

Theorem 9.4 For every even number $N \in \mathbb{N}$, let $\lambda_t^{(N)}$, $t = 0, \ldots, \frac{N}{2}$, be the eigenvalues of A_N as given in (7.1). Assume there is a sequence of even numbers $(N_n)_{n\in\mathbb{N}} \subset \mathbb{N}$ with $N_n \nearrow \infty$ for $n \to \infty$ and a sequence $(t_n)_{n\in\mathbb{N}} \subset \mathbb{N}$ with $t_n \in \{1, \ldots, \frac{N_n}{2} - 1\}$ for every $n \in \mathbb{N}$ such that

$$\limsup_{n\to\infty} \lambda_{t_n}^{(N_n)} \leq 0.$$

Define a sequence $(\varphi_n)_{n\in\mathbb{N}} \subset L^2(0,2)$ of step functions by

$$\varphi_n = \sum_{k=0}^{N_n-1} \cos\left[\frac{2\pi k t_n}{N_n}\right] \chi_{\left(\frac{2}{N_n}k, \frac{2}{N_n}(k+1)\right)} \qquad (9.12)$$

for every $n \in \mathbb{N}$. Then we find a subsequence $(\varphi_{n_k})_{k \in \mathbb{N}} \subset (\varphi_n)_{n \in \mathbb{N}}$ such that

$$\varphi_{n_k} \underset{k \to \infty}{\longrightarrow} 0 \quad in \quad L^2(0,2). \tag{9.13}$$

Proof: Let $(N_n)_{n \in \mathbb{N}} \subset \mathbb{N}$ with $N_n \nearrow \infty$ for $n \to \infty$, $(t_n)_{n \in \mathbb{N}} \subset \mathbb{N}$ as described. Due to Lemma 9.2, the sequence $(\lambda_{t_n}^{(N_n)})_{n \in \mathbb{N}}$ is bounded in $\mathbb{R}$, hence we can extract a subsequence of $(N_n)_{n \in \mathbb{N}}$ (not relabelled) such that $\lim_{n \to \infty} \lambda_{t_n}^{(N_n)} = \lambda \leq 0$. For every even number $N \in \mathbb{N}$, let $\bar{\lambda}_t^{(N)}$, $t = 0, \ldots, \frac{N}{2}$ be the eigenvalues of $\bar{A}_N$ (cf.(9.3)) as given in (9.5). Then Lemma 9.1 yields

$$\lim_{n \to \infty} \bar{\lambda}_{t_n}^{(N_n)} = \lambda \leq 0. \tag{9.14}$$

For every $n \in \mathbb{N}$, the vector $v_n = (v_n^{(0)}, \ldots, v_n^{(N_n-1)}) \in \mathbb{R}^{N_n}$ with

$$v_n^{(k)} = \sqrt{\frac{2}{N_n}} \cos\left[\frac{2\pi k t_n}{N_n}\right] \quad \text{for} \quad k = 0, \ldots, N_n - 1 \tag{9.15}$$

is an eigenvector of $\bar{A}_{N_n}$ corresponding to the eigenvalue $\bar{\lambda}_{t_n}^{(N_n)}$ (cf. (9.6)). Then $(\varphi_n)_{n \in \mathbb{N}} \subset L^2(0,2)$ as given in (9.12) satisfies

$$\varphi_n = \sum_{k=0}^{N_n-1} \cos\left[\frac{2\pi k t_n}{N_n}\right] \chi_{\left(\frac{2}{N_n}k, \frac{2}{N_n}(k+1)\right)} = \sqrt{\frac{N_n}{2}} \sum_{k=0}^{N_n-1} v_n^{(k)} \chi_{\left(\frac{2}{N_n}k, \frac{2}{N_n}(k+1)\right)}.$$

Obviously, $|\varphi_n| \leq 1$ and thus

$$\|\varphi_n\|_{L^2(0,2)} \leq \sqrt{2} \tag{9.16}$$

for every $n \in \mathbb{N}$ so that we can choose a subsequence (not relabelled) weakly converging to a function $\varphi \in L^2(0,2)$, i.e.

$$\varphi_n \underset{n \to \infty}{\longrightarrow} \varphi \quad in \quad L^2(0,2). \tag{9.17}$$

Furthermore

$$\int_0^2 \varphi_n(x)dx = \sum_{k=0}^{N_n-1} \int_{\frac{2}{N_n}k}^{\frac{2}{N_n}(k+1)} \varphi_n(x)dx = \frac{2}{N_n} \sum_{k=0}^{N_n-1} \cos\left[\frac{2\pi k t_n}{N_n}\right].$$

Since $\frac{t_n}{N_n} \notin \mathbb{Z}$ we have $1 - \exp\left[\frac{2\pi t_n i}{N_n}\right] \neq 0$ and obtain

$$\sum_{k=0}^{N_n-1} \cos\left[\frac{2\pi k t_n}{N_n}\right] = Re\left(\sum_{k=0}^{N_n-1} \exp\left[\frac{2\pi k t_n i}{N_n}\right]\right) = Re\left(\sum_{k=0}^{N_n-1} \exp\left[\frac{2\pi t_n i}{N_n}\right]^k\right)$$

$$= Re\left(\frac{1 - \left(\exp\left[\frac{2\pi t_n i}{N_n}\right]\right)^{N_n}}{1 - \exp\left[\frac{2\pi t_n i}{N_n}\right]}\right) = Re\left(\frac{1 - \exp\left[2\pi t_n i\right]}{1 - \exp\left[\frac{2\pi t_n i}{N_n}\right]}\right) = 0,$$

where we used the Euler formula $\exp(ix) = \cos(x) + i\sin(x)$ in the last step. This yields

$$\int_0^2 \varphi_n(x)dx = 0 \tag{9.18}$$

for every $n \in \mathbb{N}$, which we will need later on. Now, take an arbitrary $\psi \in C_0^\infty(0,2)$ and set

$$\psi_n = \sum_{k=0}^{N_n-1} \psi(\frac{2}{N_n}k)\chi_{(\frac{2}{N_n}k,\frac{2}{N_n}(k+1))}$$

for every $n \in \mathbb{N}$. For $x \in (\frac{2}{N_n}k, \frac{2}{N_n}(k+1))$, $k \in \{0,\ldots,N_n-1\}$, we have

$$|\psi_n(x) - \psi(x)| = |\psi(\frac{2}{N_n}k) - \psi(x)| \le C|x - \frac{2}{N_n}k| \le \frac{C}{N_n} \xrightarrow[n\to\infty]{} 0,$$

i.e. $\psi_n(x) \to \psi(x)$ for a.e. $x \in (0,2)$, and using $|\psi - \psi_n|^2 \le (2\max|\psi|)^2 < \infty$ for every $n \in \mathbb{N}$, Lebesgue's dominated convergence theorem yields

$$\psi_n \xrightarrow[n\to\infty]{} \psi \quad \text{in } L^2(0,2). \tag{9.19}$$

For every $n \in \mathbb{N}$, we define $w_n = (w_n^{(0)},\ldots,w_n^{(N_n-1)}) \in \mathbb{R}^{N_n}$ by

$$w_n^{(k)} = \sqrt{\frac{2}{N_n}}\psi(\frac{2}{N_n}k)$$

for $k = 0,\ldots,N_n-1$, and recalling (9.15) we get

$$w_n^t v_n = \sum_{i=1}^{N_n} w_n^{(i-1)}v_n^{(i-1)} = \frac{2}{N_n}\sum_{i=1}^{N_n} \psi(\frac{2}{N_n}(i-1))\cos\left[\frac{2\pi(i-1)t_n}{N_n}\right]$$

$$= \sum_{i=1}^{N_n}\int_{\frac{2}{N_n}(i-1)}^{\frac{2}{N_n}i} \psi_n(x)\varphi_n(x)dx = \int_0^2 \psi_n(x)\varphi_n(x)dx, \tag{9.20}$$

and if φ_n and ψ_n also denote the periodic extensions of φ_n, ψ_n on $\mathbb{R}$, which by definition of φ_n yields $\varphi_n(x) = 1$ for $x \in (2, 2+\frac{2}{N_n})$, we obtain

$$\sum_{i=1}^{N_n-1} w_n^{(i-1)}v_n^{(i)} + w_n^{(N_n-1)}v_n^{(0)} =$$

$$= \frac{2}{N_n}\sum_{i=1}^{N_n-1}\psi(\frac{2}{N_n}(i-1))\cos\left[\frac{2\pi it_n}{N_n}\right] + \frac{2}{N_n}\psi(\frac{2}{N_n}(N_n-1))$$

$$= \sum_{i=1}^{N_n-1}\int_{\frac{2}{N_n}(i-1)}^{\frac{2}{N_n}i}\psi_n(x)\varphi_n(x+\frac{2}{N_n})dx + \int_{2-\frac{2}{N_n}}^2\psi_n(x)\varphi_n(x+\frac{2}{N_n})dx$$

$$= \int_0^2\psi_n(x)\varphi_n(x+\frac{2}{N_n})dx,$$

and using $w_n^{(0)} = \sqrt{\frac{2}{N_n}}\psi(0) = 0$ we get

$$
\sum_{i=1}^{N_n-1} w_n^{(i)} v_n^{(i-1)} + w_n^{(0)} v_n^{(N_n-1)} = \frac{2}{N_n} \sum_{i=1}^{N_n-1} \psi(\frac{2}{N_n}i) \cos\left[\frac{2\pi(i-1)t_n}{N_n}\right]
$$

$$
= \sum_{i=1}^{N_n-1} \int_{\frac{2}{N_n}i}^{\frac{2}{N_n}(i+1)} \psi_n(x)\varphi_n(x-\frac{2}{N_n})dx
$$

$$
= \int_{\frac{2}{N_n}}^{2} \psi_n(x)\varphi_n(x-\frac{2}{N_n})dx = \int_0^2 \psi_n(x)\varphi_n(x-\frac{2}{N_n})dx,
$$

since $\psi(x) = \psi(0) = 0$ for $x \in (0, \frac{2}{N_n})$ implies $\psi_n(x) = 0$ for $x \in (0, \frac{2}{N_n})$. Then, recalling (6.11),

$$
w_n^t \bar{A}_{N_n} v_n = w_n^t B_{N_n} v_n + w_n^t C_{N_n} v_n = \sum_{i=1}^{N_n}\sum_{j=1}^{N_n} w_n^{(i-1)} v_n^{(j-1)} c_{ij}^{(N_n)} + \sum_{i=1}^{N_n}\sum_{j=1}^{N_n} w_n^{(i-1)} v_n^{(j-1)} b_{|i-j|}^{(N)}
$$

$$
= \sum_{i=1}^{N_n}\sum_{j=1}^{N_n} w_n^{(i-1)} v_n^{(j-1)} c_{ij}^{(N_n)} + b_0^{(N)} \sum_{i=1}^{N_n} w_n^{(i-1)} v_n^{(i-1)}
$$

$$
+ b_1^{(N)} \sum_{i=1}^{N_n-1} w_n^{(i-1)} v_n^{(i)} + b_1^{(N)} \sum_{i=1}^{N_n-1} w_n^{(i)} v_n^{(i-1)} + b_1^{(N)} w_n^{(N_n-1)} v_n^{(0)} + b_1^{(N)} w_n^{(0)} v_n^{(N_n-1)}
$$

$$
= \sum_{i=1}^{N_n}\sum_{j=1}^{N_n} w_n^{(i-1)} v_n^{(j-1)} c_{ij}^{(N_n)} + \int_{-1}^{1} g(y)dy \int_0^2 \psi_n(x)\varphi_n(x)dx
$$

$$
+ \frac{1}{2}\int_{-1}^{1} g(y)dy \left[\int_0^2 \psi_n(x)\varphi_n(x+\frac{2}{N_n})dx + \int_0^2 \psi_n(x)\varphi_n(x-\frac{2}{N_n})dx\right]
$$

and using (6.13), we obtain

$$
\sum_{i=1}^{N_n}\sum_{j=1}^{N_n} w_n^{(i-1)} v_n^{(j-1)} c_{ij}^{(N_n)} =
$$

$$
= \frac{2}{N_n} \sum_{i=1}^{N_n}\sum_{j=1}^{N_n} \psi(\frac{2}{N_n}(i-1)) \cos\left[\frac{2\pi(j-1)t_n}{N_n}\right]\left[-N_n \int_{\frac{2}{N_n}(i-1)}^{\frac{2}{N_n}i} \int_{\frac{2}{N_n}(j-1)}^{\frac{2}{N_n}j} g(x-y)dydx\right]
$$

$$
= -2 \sum_{i=1}^{N_n}\sum_{j=1}^{N_n} \int_{\frac{2}{N_n}(i-1)}^{\frac{2}{N_n}i} \int_{\frac{2}{N_n}(j-1)}^{\frac{2}{N_n}j} g(x-y)\psi_n(x)\varphi_n(y)dydx
$$

$$
= -2 \int_0^2 \int_0^2 g(x-y)\psi_n(x)\varphi_n(y)dydx .
$$

Combining these calculations we can rewrite $w_n^t \bar{A}_{N_n} v_n$ in the form

$$
w_n^t \bar{A}_{N_n} v_n = \int_{-1}^{1} g(y)dy \int_0^2 \psi_n(x)\left[\varphi_n(x) + \frac{1}{2}\varphi_n(x+\frac{2}{N_n}) + \frac{1}{2}\varphi_n(x-\frac{2}{N_n})\right]dx
$$

$$-2 \int_0^2 \int_0^2 g(x-y) \psi_n(x) \varphi_n(y) dy dx$$
$$=: \mathcal{B}_n(\psi_n, \varphi_n).$$

Since v_n is an eigenvector of $\bar{A}_{N_n}$ corresponding to the eigenvalue $\bar{\lambda}_{t_n}^{(N_n)}$, we derive from (9.20)

$$\bar{\lambda}_{t_n}^{(N_n)} \int_0^2 \psi_n(x) \varphi_n(x) dx = \bar{\lambda}_{t_n}^{(N_n)} w_n^t v_n = w_n^t \bar{A}_{N_n} v_n = \mathcal{B}_n(\psi_n, \varphi_n). \tag{9.21}$$

We want to show that

$$\lim_{n \to \infty} \mathcal{B}_n(\psi_n, \varphi_n) = \mathcal{B}(\psi, \varphi),$$

where $\mathcal{B} : L^2(0,2) \times L^2(0,2) \to \mathbb{R}$ is the symmetric, continuous bilinear form given by (9.11). To do so, we observe that

$$\begin{aligned}
\mathcal{B}_n(\psi_n, \varphi_n) - \mathcal{B}(\psi, \varphi) &= \int_{-1}^1 g(y) dy \int_0^2 (\psi_n(x) \varphi_n(x) - \psi(x) \varphi(x)) dx \\
&\quad + \frac{1}{2} \int_{-1}^1 g(y) dy \int_0^2 (\psi_n(x) \varphi_n(x + \frac{2}{N_n}) - \psi(x) \varphi(x)) dx \\
&\quad + \frac{1}{2} \int_{-1}^1 g(y) dy \int_0^2 (\psi_n(x) \varphi_n(x - \frac{2}{N_n}) - \psi(x) \varphi(x)) dx \\
&\quad + 2 \int_0^2 \int_0^2 g(x-y)(\psi(x) \varphi(y) - \psi_n(x) \varphi_n(y)) dy dx \\
&=: I_n^{(1)} + I_n^{(2)} + I_n^{(3)} + I_n^{(4)},
\end{aligned} \tag{9.22}$$

where we can estimate $I_n^{(1)}$ as follows:

$$\begin{aligned}
|I_n^{(1)}| &\leq \int_{-1}^1 g(y) dy |(\psi_n - \psi, \varphi_n)_{L^2(0,2)} + (\psi, \varphi_n - \varphi)_{L^2(0,2)}| \\
&\leq C \left(\|\psi_n - \psi\|_{L^2(0,2)} \|\varphi_n\|_{L^2(0,2)} + |(\psi, \varphi_n - \varphi)_{L^2(0,2)}| \right) \\
&\leq C \left(\|\psi_n - \psi\|_{L^2(0,2)} |(\psi, \varphi_n - \varphi)_{L^2(0,2)}| \right),
\end{aligned}$$

where we used the boundedness of g, the Cauchy-Schwarz inequality and (9.16). Application of (9.17) and (9.19) immediately yields

$$\lim_{n \to \infty} |I_n^{(1)}| = 0.$$

As for $I_n^{(2)}$, using the boundedness of g, the Cauchy-Schwarz inequality, the 2-periodicity of φ_n, φ and ψ on $\mathbb{R}$ and the boundedness of $(\varphi_n)_{n \in \mathbb{N}}$ and $(\psi_n)_{n \in \mathbb{N}}$ (cf. (9.16), (9.19)) in $L^2(0,2)$, we deduce

$$\begin{aligned}
|I_n^{(2)}| &\leq C |(\psi_n, \varphi_n(\cdot + \frac{2}{N_n}))_{L^2(0,2)} - (\psi, \varphi)_{L^2(0,2)}| \\
&\leq C |(\psi_n - \psi, \varphi_n(\cdot + \frac{2}{N_n}))_{L^2(0,2)}| + C |(\psi, \varphi_n(\cdot + \frac{2}{N_n}) - \varphi(\cdot + \frac{2}{N_n}))_{L^2(0,2)}|
\end{aligned}$$

$$+ C |(\psi, \varphi(\cdot + \frac{2}{N_n}) - \varphi)_{L^2(0,2)}|$$

$$\leq C \|\psi_n - \psi\|_{L^2(0,2)} \|\varphi_n(\cdot + \frac{2}{N_n})\|_{L^2(0,2)} + C |(\psi(\cdot - \frac{2}{N_n}), \varphi_n - \varphi)_{L^2(0,2)}|$$

$$+ C \|\psi\|_{L^2(0,2)} \|\varphi(\cdot + \frac{2}{N_n}) - \varphi\|_{L^2(0,2)}$$

$$\leq C \|\psi_n - \psi\|_{L^2(0,2)} \|\varphi_n\|_{L^2(0,2)} + C |(\psi(\cdot - \frac{2}{N_n}) - \psi, \varphi_n - \varphi)_{L^2(0,2)}|$$

$$+ C |(\psi, \varphi_n - \varphi)_{L^2(0,2)}| + C \|\psi\|_{L^2(0,2)} \|\varphi(\cdot + \frac{2}{N_n}) - \varphi\|_{L^2(0,2)}$$

$$\leq C \|\psi_n - \psi\|_{L^2(0,2)} + C \|\psi(\cdot - \frac{2}{N_n}) - \psi\|_{L^2(0,2)} \|\varphi_n - \varphi\|_{L^2(0,2)}$$

$$+ C |(\psi, \varphi_n - \varphi)_{L^2(0,2)}| + C \|\varphi(\cdot + \frac{2}{N_n}) - \varphi\|_{L^2(0,2)}$$

$$\leq C \|\psi_n - \psi\|_{L^2(0,2)} + C |(\psi, \varphi_n - \varphi)_{L^2(0,2)}|$$

$$+ C \|\varphi(\cdot + \frac{2}{N_n}) - \varphi\|_{L^2(0,2)} + C \|\psi(\cdot + \frac{2}{N_n}) - \psi\|_{L^2(0,2)} .$$

Using (9.19) and (9.17), we conclude that the first two terms converge to 0 as $n \to \infty$. Since $N_n \to \infty$ for $n \to \infty$, the other terms also converge to 0 as $n \to \infty$, and we infer

$$\lim_{n\to\infty} |I_n^{(2)}| = 0 .$$

Analogously, we prove that

$$\lim_{n\to\infty} |I_n^{(3)}| = 0 .$$

Finally, using the boundedness of g,

$$|I_n^{(4)}| \leq \left| \int_0^2 \int_0^2 g(x-y)\psi(x)(\varphi(y) - \varphi_n(y)) dy dx \right|$$

$$+ \left| \int_0^2 \int_0^2 g(x-y)(\psi(x) - \psi_n(x))\varphi_n(y) dy dx \right|$$

$$\leq \left| \int_0^2 \int_0^2 g(x-y)\psi(x)(\varphi(y) - \varphi_n(y)) dx dy \right|$$

$$+ C \int_0^2 |\psi(x) - \psi_n(x)| dx \int_0^2 |\varphi_n(y)| dy$$

$$= |(T, \varphi - \varphi_n)_{L^2(0,2)}| + C \|\psi - \psi_n\|_{L^1(0,2)} \|\varphi_n\|_{L^1(0,2)} ,$$

where $T : (0,2) \to \mathbb{R}$ is defined as

$$T(y) = \int_0^2 g(x-y)\psi(x) dx .$$

Due to the boundedness of g, we obtain the pointwise estimate $|T| \leq C \|\psi\|_{L^1(0,2)} \leq C \|\psi\|_{L^2(0,2)} \leq C$. Thus T is bounded, which implies $T \in L^2(0,2)$, and using (9.17) we deduce

$$\lim_{n\to\infty} |(T, \varphi - \varphi_n)_{L^2(0,2)}| = 0 .$$

Furthermore, due to the boundedness (9.16)

$$\|\psi - \psi_n\|_{L^1(0,2)} \|\varphi_n\|_{L^1(0,2)} \le C \|\psi - \psi_n\|_{L^2(0,2)} \to 0$$

for $n \to \infty$ (cf. (9.19)), and we conclude

$$\lim_{n\to\infty} |I_n^{(4)}| = 0 \,.$$

From (9.22), we conclude

$$\lim_{n\to\infty} \mathcal{B}_n(\psi_n, \varphi_n) = \mathcal{B}(\psi, \varphi) \,. \tag{9.23}$$

Consider now the left-hand side of (9.21), which can be written as

$$\bar{\lambda}_{t_n}^{(N_n)} \int_0^2 \psi_n(x) \varphi_n(x) dx = \bar{\lambda}_{t_n}^{(N_n)} (\psi_n, \varphi_n)_{L^2(0,2)} = \bar{\lambda}_{t_n}^{(N_n)} [(\psi_n - \psi, \varphi_n)_{L^2(0,2)} + (\psi, \varphi_n)_{L^2(0,2)}] \,. \tag{9.24}$$

Using the Cauchy-Schwarz inequality and (9.19), we deduce

$$|(\psi_n - \psi, \varphi_n)_{L^2(0,2)}| \le \|\psi_n - \psi\|_{L^2(0,2)} \|\varphi_n\|_{L^2(0,2)} \le C \|\psi_n - \psi\|_{L^2(0,2)} \,,$$

which together with (9.19) yields

$$\lim_{n\to\infty} (\psi_n - \psi, \varphi_n)_{L^2(0,2)} = 0 \,.$$

From the weak convergence property (9.17), we obtain

$$\lim_{n\to\infty} (\psi, \varphi_n)_{L^2(0,2)} = (\psi, \varphi)_{L^2(0,2)} \,,$$

and recalling (9.14), we deduce from (9.24)

$$\lim_{n\to\infty} \bar{\lambda}_{t_n}^{(N_n)} \int_0^2 \psi_n(x) \varphi_n(x) dx = \lambda(\psi, \varphi)_{L^2(0,2)} \,,$$

which together with (9.23) and (9.21) implies

$$\mathcal{B}(\psi, \varphi) = \lambda(\psi, \varphi)_{L^2(0,2)} \,. \tag{9.25}$$

Since $\psi \in C_0^\infty(0,2)$ was chosen arbitrarily, (9.25) holds for every $\psi \in C_0^\infty(0,2)$. Since $C_0^\infty(0,2)$ is dense in $L^2(0,2)$ and because of the $L^2(0,2)$-continuity of both-hand sides of (9.25) (the one of $\mathcal{B}$ following from Lemma 9.3), (9.25) holds for every $\psi \in L^2(0,2)$. Choosing $\psi = \varphi$ we obtain, using the 2-periodicity of g on $\mathbb{R}$ and the symmetry of g,

$$\begin{aligned}
\lambda \|\varphi\|_{L^2(0,2)}^2 &= \mathcal{B}(\varphi, \varphi) = 2 \int_{-1}^1 g(y) dy \int_0^2 \varphi(x)^2 dx - 2 \int_0^2 \int_0^2 g(x-y) \varphi(x) \varphi(y) dy dx \\
&= 2 \int_0^2 \int_0^2 g(y-x) \varphi(x)^2 dy dx - 2 \int_0^2 \int_0^2 g(x-y) \varphi(x) \varphi(y) dy dx \\
&= \int_0^2 \int_0^2 g(x-y)(\varphi(x)^2 - 2\varphi(x)\varphi(y)) dy dx + \int_0^2 \int_0^2 g(y-x) \varphi(x)^2 dx dy
\end{aligned}$$

$$= \int_0^2 \int_0^2 g(x-y)(\varphi(x)^2 - 2\varphi(x)\varphi(y) + \varphi(y)^2)dydx$$

$$= \int_0^2 \int_0^2 g(x-y)(\varphi(x) - \varphi(y))^2 dydx.$$

(9.18) and (9.17) immediately imply

$$\int_0^2 \varphi(x)dx = \lim_{n\to\infty} \int_0^2 \varphi_n(x)dx = 0.$$

Due to (9.14) and g being strictly positive (i.e. $g \geq g_0 > 0$ on $\mathbb{R}$, see conditions on g in Chapter 2), we get

$$0 \geq \lambda\|\varphi\|_{L^2(0,2)}^2 \geq g_0 \int_0^2 \int_0^2 (\varphi(x) - \varphi(y))^2 dydx$$

$$= 4g_0 \int_0^2 \varphi(x)^2 dx - 2g_0 \int_0^2 \varphi(x)dx \int_0^2 \varphi(y)dy = 4g_0\|\varphi\|_{L^2(0,2)}^2,$$

which, since $g_0 > 0$, can only hold if $\varphi = 0$, and due to (9.17), this implies (9.13). $\qquad\square$

Theorem 9.5 *For every even number $N \in \mathbb{N}$, let $\lambda_t^{(N)}$, $t = 0, \dots, \frac{N}{2}$, be the eigenvalues of A_N as given in (7.1). Assume there is a sequence of even numbers $(N_n)_{n\in\mathbb{N}} \subset \mathbb{N}$ with $N_n \nearrow \infty$ for $n \to \infty$ and a sequence $(t_n)_{n\in\mathbb{N}} \subset \mathbb{N}$ with $t_n \in \{1, \dots, \frac{N_n}{2} - 1\}$ for every $n \in \mathbb{N}$ such that*

$$\limsup_{n\to\infty} \lambda_{t_n}^{(N_n)} \leq 0.$$

Then we find a subsequence of $(N_n)_{n\in\mathbb{N}}$ (not relabelled) such that

$$\lim_{n\to\infty} \lambda_{t_n}^{(N_n)} = 0, \quad \lim_{n\to\infty} \frac{t_n}{N_n} = \frac{1}{2}.$$

Proof: Let $(N_n)_{n\in\mathbb{N}}$, $N_n \nearrow \infty$ such that for every $n \in \mathbb{N}$, we find $t_n \in \{1, \dots, \frac{N_n}{2} - 1\}$ with $\limsup_{n\to\infty} \lambda_{t_n}^{(N_n)} \leq 0$. Using Lemma 9.2, we can choose a subsequence of $(\lambda_{t_n}^{(N_n)})_{n\in\mathbb{N}}$ (not relabelled) such that

$$\lim_{n\to\infty} \lambda_{t_n}^{(N_n)} = \lambda \leq 0. \tag{9.26}$$

For every even $N \in \mathbb{N}$, let $\bar{\lambda}_t^{(N)}$, $t \in \{0, \dots, \frac{N}{2}\}$ (cf. (7.1)), be the eigenvalues of the matrix $\bar{A}_N \in \mathbb{R}^{N\times N}$ given by (9.3). From Lemma 9.1, we know that

$$\left|\lambda_{t_n}^{(N_n)} - \bar{\lambda}_{t_n}^{(N_n)}\right| \leq \frac{C}{N_n} \tag{9.27}$$

for every $n \in \mathbb{N}$. Furthermore, for every $n \in \mathbb{N}$ we have

$$\bar{\lambda}_{t_n}^{(N_n)} = \bar{a}_0^{(N_n)} + (-1)^{t_n}\bar{a}_{\frac{N_n}{2}}^{(N_n)} + 2\sum_{k=1}^{\frac{N_n}{2}-1} \bar{a}_k^{(N_n)} \cos\left[\frac{2\pi k t_n}{N_n}\right]$$

$$
\begin{aligned}
= \ & b_0^{(N)} + 2b_1^{(N)} \cos\left[\frac{2\pi t_n}{N_n}\right] + c_0^{(N_n)} + (-1)^{t_n} c_{\frac{N_n}{2}}^{(N_n)} + 2 \sum_{k=1}^{\frac{N_n}{2}-1} c_k^{(N_n)} \cos\left[\frac{2\pi k t_n}{N_n}\right] \\
= \ & \int_0^2 g(y)dy \left(1 + \cos\left[\frac{2\pi t_n}{N_n}\right]\right) \\
& + \sum_{k=0}^{\frac{N_n}{2}-1} c_k^{(N_n)} \cos\left[\frac{2\pi k t_n}{N_n}\right] + \sum_{k=1}^{\frac{N_n}{2}} c_k^{(N_n)} \cos\left[\frac{2\pi k t_n}{N_n}\right] .
\end{aligned}
$$

For every $n \in \mathbb{N}$, define $\tilde{c}_k^{(N_n)}$, $k = 0, \dots, \frac{N_n}{2}$ by

$$
\tilde{c}_k^{(N_n)} = -2 \int_{\frac{2}{N_n}k}^{\frac{2}{N_n}(k+1)} g(y)dy .
$$

In the definition of $c_k^{(N)}$ (cf. (6.12)), we may replace $g(x-y)$ by $g(y-x)$ due to the symmetry of g. Using the Lipschitz continuity of g, we thus obtain

$$
\begin{aligned}
\left|\tilde{c}_k^{(N_n)} - c_k^{(N_n)}\right| &= \left| N_n \int_0^{\frac{2}{N_n}} \int_{\frac{2}{N_n}k}^{\frac{2}{N_n}(k+1)} g(y-x)dydx - N_n \int_0^{\frac{2}{N_n}} \int_{\frac{2}{N_n}k}^{\frac{2}{N_n}(k+1)} g(y)dydx \right| \\
&\leq N_n \int_0^{\frac{2}{N_n}} \int_{\frac{2}{N_n}k}^{\frac{2}{N_n}(k+1)} |g(y-x) - g(y)|dydx \\
&\leq CN_n \int_0^{\frac{2}{N_n}} \int_{\frac{2}{N_n}k}^{\frac{2}{N_n}(k+1)} |x|dydx \leq C \int_0^{\frac{2}{N_n}} \int_{\frac{2}{N_n}k}^{\frac{2}{N_n}(k+1)} dydx \leq \frac{C}{N_n^2} .
\end{aligned}
$$

This implies that if we define

$$
\begin{aligned}
\mu_{t_n}^{(N_n)} = \ & \int_0^2 g(y)dy \left(1 + \cos\left[\frac{2\pi t_n}{N_n}\right]\right) \\
& + \sum_{k=0}^{\frac{N_n}{2}-1} \tilde{c}_k^{(N_n)} \cos\left[\frac{2\pi k t_n}{N_n}\right] + \sum_{k=1}^{\frac{N_n}{2}} \tilde{c}_k^{(N_n)} \cos\left[\frac{2\pi k t_n}{N_n}\right]
\end{aligned} \tag{9.28}
$$

for all $n \in \mathbb{N}$, we can compare this term with $\bar{\lambda}_{t_n}^{(N_n)}$ as follows:

$$
\begin{aligned}
\left|\bar{\lambda}_{t_n}^{(N_n)} - \mu_{t_n}^{(N_n)}\right| &= \left| \sum_{k=0}^{\frac{N_n}{2}-1} (c_k^{(N_n)} - \tilde{c}_k^{(N_n)}) \cos\left[\frac{2\pi k t_n}{N_n}\right] + \sum_{k=1}^{\frac{N_n}{2}} (c_k^{(N_n)} - \tilde{c}_k^{(N_n)}) \cos\left[\frac{2\pi k t_n}{N_n}\right] \right| \\
&\leq \sum_{k=0}^{\frac{N_n}{2}-1} |c_k^{(N_n)} - \tilde{c}_k^{(N_n)}| + \sum_{k=1}^{\frac{N_n}{2}} |c_k^{(N_n)} - \tilde{c}_k^{(N_n)}| \leq \sum_{k=0}^{\frac{N_n}{2}-1} \frac{C}{N_n^2} + \sum_{k=1}^{\frac{N_n}{2}} \frac{C}{N_n^2} \leq \frac{C}{N_n} .
\end{aligned}
$$

Using (9.27), we thus obtain

$$
\lambda_{t_n}^{(N_n)} \geq \bar{\lambda}_{t_n}^{(N_n)} - \frac{C}{N_n} \geq \mu_{t_n}^{(N_n)} - \frac{C}{N_n} \tag{9.29}
$$

for every $n \in \mathbb{N}$. We define a sequence of functions $(\varphi_n)_{n \in \mathbb{N}} \subset L^2(0, 2)$ as in (9.12). Then, using Theorem 9.4, we may assume that $\varphi_n \longrightarrow_{n \to \infty} 0$ in $L^2(0, 2)$. Furthermore, we have

$$\sum_{k=0}^{\frac{N_n}{2}-1} \tilde{c}_k^{(N_n)} \cos \left[\frac{2\pi k t_n}{N_n} \right] = -2 \sum_{k=0}^{\frac{N_n}{2}-1} \int_{\frac{2}{N_n}k}^{\frac{2}{N_n}(k+1)} g(y) dy \cos \left[\frac{2\pi k t_n}{N_n} \right] = -2 \int_0^1 g(y)\varphi_n(y) dy,$$

$$\sum_{k=1}^{\frac{N_n}{2}} \tilde{c}_k^{(N_n)} \cos \left[\frac{2\pi k t_n}{N_n} \right] = -2 \sum_{k=1}^{\frac{N_n}{2}} \int_{\frac{2}{N_n}k}^{\frac{2}{N_n}(k+1)} g(y) dy \cos \left[\frac{2\pi k t_n}{N_n} \right] = -2 \int_{\frac{2}{N_n}}^{1+\frac{2}{N_n}} g(y)\varphi_n(y) dy,$$

so that we can rewrite $\mu_{t_n}^{(N_n)}$ as

$$\begin{aligned}
\mu_{t_n}^{(N_n)} &= \int_0^2 g(y) dy \left(1 + \cos \left[\frac{2\pi t_n}{N_n} \right] \right) - 2 \int_0^1 g(y)\varphi_n(y) dy - 2 \int_{\frac{2}{N_n}}^{1+\frac{2}{N_n}} g(y)\varphi_n(y) dy \\
&= \int_0^2 g(y) dy \left(1 + \cos \left[\frac{2\pi t_n}{N_n} \right] \right) - 4 \int_0^1 g(y)\varphi_n(y) dy + R_n,
\end{aligned}$$

where R_n can be estimated as follows, using $|\varphi_n| \leq 1$ and the boundedness of g:

$$|R_n| = \left| 2 \int_0^{\frac{2}{N_n}} g(y)\varphi_n(y) dy - 2 \int_1^{1+\frac{2}{N_n}} g(y)\varphi_n(y) dy \right| \leq \frac{C}{N_n}.$$

Combining this with (9.29), we obtain

$$\lambda_{t_n}^{(N_n)} \geq \int_0^2 g(y) dy \left(1 + \cos \left[\frac{2\pi t_n}{N_n} \right] \right) - 4 \int_0^1 g(y)\varphi_n(y) dy - \frac{C}{N_n}. \tag{9.30}$$

Since $g \cdot \chi_{(0,1)} \in L^2(0, 2)$ and $\varphi_n \longrightarrow_{n \to \infty} 0$ in $L^2(0, 2)$, we have

$$\lim_{n \to \infty} \int_0^1 g(y)\varphi_n(y) dy = \lim_{n \to \infty} \int_0^2 (g \cdot \chi_{(0,1)})(y)\varphi_n(y) dy = 0.$$

Due to (9.26), taking the limit for $n \to \infty$ in (9.30) yields

$$0 \geq \lambda \geq \int_0^2 g(y) dy \cdot \limsup_{n \to \infty} \left(1 + \cos \left[\frac{2\pi t_n}{N_n} \right] \right) \geq 0$$

$$0 \geq \lambda \geq \int_0^2 g(y) dy \cdot \liminf_{n \to \infty} \left(1 + \cos \left[\frac{2\pi t_n}{N_n} \right] \right) \geq 0$$

which implies $\lambda = 0$. Furthermore, since $\int_0^2 g(y) dy > 0$, we deduce

$$\lim_{n \to \infty} \left(1 + \cos \left[\frac{2\pi t_n}{N_n} \right] \right) = 0,$$

which due to $\frac{2\pi t_n}{N_n} \in [0, \pi]$ implies that

$$\lim_{n \to \infty} \frac{t_n}{N_n} = \frac{1}{2},$$

and the proof is complete. $\qquad\square$

An easy consequence of Theorem 9.5 is the following:

Corollary 9.6 *For every even number $N \in \mathbb{N}$, let $\lambda_t^{(N)}$, $t = 0, \ldots, \frac{N}{2}$, be the eigenvalues of A_N as given in (7.1). Then for every $\delta > 0$, we find $\nu_0 > 0$ and $N_0 \in \mathbb{N}$ such that for every even number $N \geq N_0$, the following condition holds:*

$$\lambda_t^{(N)} \geq \nu_0 \quad \text{for every} \quad t \in \{1, \ldots, \frac{N}{2} - 1\} \quad \text{satisfying} \quad \frac{t}{N} \leq \frac{1}{2} - \delta. \tag{9.31}$$

Proof: Assume that this is not the case. Then there exists a $\delta_0 > 0$ such that for every $n \in \mathbb{N}$ we find an even number $N_n \geq n$ so that (9.31) is not satisfied for $\nu_0 = \frac{1}{n}$ and $N = N_n$. This means that for every $n \in \mathbb{N}$, we find $t_n \in \{1, \ldots, \frac{N_n}{2} - 1\}$ satisfying $\frac{t_n}{N_n} \leq \frac{1}{2} - \delta_0$ such that $\lambda_{t_n}^{(N_n)} \leq \frac{1}{n}$. In particular,

$$\limsup_{n \to \infty} \lambda_{t_n}^{(N_n)} \leq 0.$$

Since $N_n \geq n$, the condition $N_n \nearrow \infty$ for $n \to \infty$ is also fulfilled, so that we can apply Theorem 9.5 to obtain a subsequence of $(N_n)_{n \in \mathbb{N}}$ (not relabelled) that satisfies

$$\lim_{n \to \infty} \frac{t_n}{N_n} = \frac{1}{2},$$

which contradicts $\frac{t_n}{N_n} \leq \frac{1}{2} - \delta_0$. $\qquad\qquad\square$

Thus, to obtain positivity of the eigenvalues $\lambda_t^{(N)}$, $t = 1, \ldots, \frac{N}{2} - 1$ for large N, we only have to consider indices t "close" to $\frac{N}{2}$ in the sense that $\frac{t}{N} \approx \frac{1}{2}$. This will be done in the next chapter.

Chapter 10

Estimation of the Eigenvalues

10.1 Basic Computations

In the previous chapter, we have shown that for large even N, λ_t is positive if $t \in \{1, \ldots, \frac{N}{2} - 1\}$ with $\frac{t}{N} \leq \frac{1}{2} - \delta$, δ sufficiently small. In order to show that all $\lambda_t^{(N)}$, $t \in \{1, \ldots, \frac{N}{2} - 1\}$ are positive (at least if N is large), we will have to estimate $\lambda_t^{(N)}$ for t "close to" $\frac{N}{2}$ in that sense that $\frac{t}{N} > \frac{1}{2} - \delta$. The idea behind these estimates is to introduce a parameter $\delta_{t,N}$ defined by

$$\delta_{t,N} = 1 - \frac{2t}{N}$$

for $t \in \{1, \ldots, \frac{N}{2} - 1\}$ and N even and to estimate the eigenvalues $\lambda_t^{(N)}$ for indices t for which $\delta_{t,N}$ is small in terms of $\delta_{t,N}$ and N to finally conclude positivity. We recall that

$$\lambda_t^{(N)} = a_0^{(N)} + (-1)^t a_{\frac{N}{2}}^{(N)} + 2 \sum_{k=1}^{\frac{N}{2}-1} a_k^{(N)} \cos\left[\frac{2\pi k t}{N}\right] \quad \text{for} \quad t = 1, \ldots, \frac{N}{2} - 1, \qquad (10.1)$$

where by Theorem 6.4,

$$
\begin{aligned}
a_0^{(N)} &= b_0^{(N)} + c_0^{(N)} + d_0^{(N)} \\
a_1^{(N)} = a_{N-1}^{(N)} &= \tfrac{1}{2} b_0^{(N)} + c_1^{(N)} - \tfrac{1}{2} d_0^{(N)} \\
a_k^{(N)} &= c_k^{(N)} \qquad \text{for} \quad k = 2, \ldots, N-2
\end{aligned}
$$

with

$$b_0^{(N)} = \int_{-1}^{1} g(y)dy, \quad d_0^{(N)} = N(-1)^{\frac{N}{2}+1} \int_{-1}^{1} g(y)\bar{u}_N(y)dy$$

$$c_k^{(N)} = -N \int_0^{\frac{2}{N}} \int_{\frac{2}{N}k}^{\frac{2}{N}(k+1)} g(x-y)dxdy$$

for $k = 0, \ldots, N-1$, hence

$$\lambda_t^{(N)} = b_0^{(N)} \left(1 + \cos\left[\frac{2\pi t}{N}\right]\right) + d_0^{(N)} \left(1 - \cos\left[\frac{2\pi t}{N}\right]\right)$$

$$+ c_0^{(N)} + (-1)^t c_{\frac{N}{2}}^{(N)} + 2 \sum_{k=1}^{\frac{N}{2}-1} c_k^{(N)} \cos\left[\frac{2\pi kt}{N}\right]$$

$$= \int_{-1}^{1} g(y)dy \left(1 + \cos\left[\frac{2\pi t}{N}\right]\right) + N(-1)^{\frac{N}{2}+1} \int_{-1}^{1} g(y)\bar{u}_N(y)dy \left(1 - \cos\left[\frac{2\pi t}{N}\right]\right)$$

$$+ c_0^{(N)} + (-1)^t c_{\frac{N}{2}}^{(N)} + 2 \sum_{k=1}^{\frac{N}{2}-1} c_k^{(N)} \cos\left[\frac{2\pi kt}{N}\right]. \tag{10.2}$$

We have to tweak this formula a little to do the estimations. Define $G : [0,1] \to \mathbb{R}$ by

$$G(x) = \int_0^x g(\xi)d\xi,$$

then $G' = g$ on $[0,1]$ and $G(0) = 0$. Using the symmetry of both $\bar{u}_N$ and g and recalling $\bar{u}_N(1) = -\frac{1}{N}$ (see Lemma 6.1), we obtain

$$\begin{aligned}
\int_{-1}^{1} g(y)\bar{u}_N(y)dy &= 2\int_0^1 g(y)\bar{u}_N(y)dy = 2\int_0^1 G'(y)\bar{u}_N(y)dy \\
&= -2\int_0^1 G(y)\bar{u}_N'(y)dy + 2G(1)\bar{u}_N(1) - 2G(0)\bar{u}_N(0) \\
&= -2\sum_{k=1}^{\frac{N}{2}} \int_{\frac{2}{N}(k-1)}^{\frac{2}{N}k} G(y)\bar{u}_N'(y)dy - \frac{2}{N}G(1).
\end{aligned}$$

Since by (6.1), $\bar{t}_{k+\frac{N}{2}} = \frac{2}{N}k$ for $k = 0,\ldots,\frac{N}{2}$, we deduce from (6.3), (6.2) that $\bar{u}_N'(x) = (-1)^{k+\frac{N}{2}-1}$ for $x \in (\frac{2}{N}(k-1), \frac{2}{N}k)$, $k = 1,\ldots,\frac{N}{2}$, so that

$$\begin{aligned}
\int_{-1}^{1} g(y)\bar{u}_N(y)dy &= 2\sum_{k=1}^{\frac{N}{2}} (-1)^{k+\frac{N}{2}} \int_{\frac{2}{N}(k-1)}^{\frac{2}{N}k} G(y)dy - \frac{2}{N}\int_0^1 g(y)dy \\
&= 2(-1)^{\frac{N}{2}} \sum_{k=0}^{\frac{N}{2}-1} (-1)^{k+1} \int_{\frac{2}{N}k}^{\frac{2}{N}(k+1)} G(y)dy - \frac{2}{N}\int_0^1 g(y)dy
\end{aligned}$$

and thus

$$N(-1)^{\frac{N}{2}+1} \int_{-1}^{1} g(y)\bar{u}_N(y)dy = 2N \sum_{k=0}^{\frac{N}{2}-1} (-1)^k \int_{\frac{2}{N}k}^{\frac{2}{N}(k+1)} G(y)dy + 2(-1)^{\frac{N}{2}} \int_0^1 g(y)dy. \tag{10.3}$$

As for the sum, by definition of G we have

$$\sum_{k=0}^{\frac{N}{2}-1} (-1)^k \int_{\frac{2}{N}k}^{\frac{2}{N}(k+1)} G(y)dy = \sum_{k=0}^{\frac{N}{2}-1} (-1)^k \int_{\frac{2}{N}k}^{\frac{2}{N}(k+1)} \left[\int_0^{\frac{2}{N}k} g(x)dx + \int_{\frac{2}{N}k}^{y} g(x)dx\right] dy$$

$$= \frac{2}{N} \sum_{k=1}^{\frac{N}{2}-1} (-1)^k \int_0^{\frac{2}{N}k} g(x)dx + \sum_{k=0}^{\frac{N}{2}-1} (-1)^k \int_{\frac{2}{N}k}^{\frac{2}{N}(k+1)} \int_{\frac{2}{N}k}^{y} g(x)dxdy. \qquad (10.4)$$

If $\frac{N}{2}$ is odd, then

$$\sum_{k=1}^{\frac{N}{2}-1} (-1)^k \int_0^{\frac{2}{N}k} g(x)dx = \sum_{\substack{k=1 \\ k \text{ odd}}}^{\frac{N}{2}-1} \left[\int_0^{\frac{2}{N}(k+1)} g(x)dx - \int_0^{\frac{2}{N}k} g(x)dx \right] = \sum_{\substack{k=1 \\ k \text{ odd}}}^{\frac{N}{2}-1} \int_{\frac{2}{N}k}^{\frac{2}{N}(k+1)} g(x)dx,$$

and if $\frac{N}{2}$ is even, we have

$$\sum_{k=1}^{\frac{N}{2}-1} (-1)^k \int_0^{\frac{2}{N}k} g(x)dx = \sum_{k=1}^{\frac{N}{2}} (-1)^k \int_0^{\frac{2}{N}k} g(x)dx - \int_0^1 g(x)dx$$

$$= \sum_{\substack{k=1 \\ k \text{ odd}}}^{\frac{N}{2}-1} \left[\int_0^{\frac{2}{N}(k+1)} g(x)dx - \int_0^{\frac{2}{N}k} g(x)dx \right] - \int_0^1 g(x)dx$$

$$= \sum_{\substack{k=1 \\ k \text{ odd}}}^{\frac{N}{2}-1} \int_{\frac{2}{N}k}^{\frac{2}{N}(k+1)} g(x)dx - \int_0^1 g(x)dx$$

and thus

$$\sum_{k=1}^{\frac{N}{2}-1} (-1)^k \int_0^{\frac{2}{N}k} g(x)dx = \sum_{\substack{k=1 \\ k \text{ odd}}}^{\frac{N}{2}-1} \int_{\frac{2}{N}k}^{\frac{2}{N}(k+1)} g(x)dx - \frac{1}{2}(1 + (-1)^{\frac{N}{2}}) \int_0^1 g(x)dx$$

in all cases, which, set into (10.4), yields

$$\sum_{k=0}^{\frac{N}{2}-1} (-1)^k \int_{\frac{2}{N}k}^{\frac{2}{N}(k+1)} G(y)dy = \frac{2}{N} \sum_{\substack{k=1 \\ k \text{ odd}}}^{\frac{N}{2}-1} \int_{\frac{2}{N}k}^{\frac{2}{N}(k+1)} g(x)dx - \frac{1}{N}(1 + (-1)^{\frac{N}{2}}) \int_0^1 g(x)dx$$

$$+ \sum_{k=0}^{\frac{N}{2}-1} (-1)^k \int_{\frac{2}{N}k}^{\frac{2}{N}(k+1)} \int_{\frac{2}{N}k}^{y} g(x)dxdy.$$

(10.3) now delivers

$$N(-1)^{\frac{N}{2}+1} \int_{-1}^{1} g(y)\bar{u}_N(y)dy = 4 \sum_{\substack{k=1 \\ k \text{ odd}}}^{\frac{N}{2}-1} \int_{\frac{2}{N}k}^{\frac{2}{N}(k+1)} g(x)dx - 2(1 + (-1)^{\frac{N}{2}}) \int_0^1 g(x)dx$$

$$+ 2N \sum_{k=0}^{\frac{N}{2}-1} (-1)^k \int_{\frac{2}{N}k}^{\frac{2}{N}(k+1)} \int_{\frac{2}{N}k}^{y} g(x)dxdy + 2(-1)^{\frac{N}{2}} \int_0^1 g(x)dx$$

$$= 4 \sum_{\substack{k=1 \\ k \text{ odd}}}^{\frac{N}{2}-1} \int_{\frac{2}{N}k}^{\frac{2}{N}(k+1)} g(x)dx - 2 \sum_{k=0}^{\frac{N}{2}-1} \int_{\frac{2}{N}k}^{\frac{2}{N}(k+1)} g(x)dx + 2N \sum_{k=0}^{\frac{N}{2}-1} (-1)^k \int_{\frac{2}{N}k}^{\frac{2}{N}(k+1)} \int_{\frac{2}{N}k}^{y} g(x)dxdy$$

$$= 2 \sum_{\substack{k=1 \\ k \text{ odd}}}^{\frac{N}{2}-1} \int_{\frac{2}{N}k}^{\frac{2}{N}(k+1)} g(x)dx - 2 \sum_{\substack{k=0 \\ k \text{ even}}}^{\frac{N}{2}-1} \int_{\frac{2}{N}k}^{\frac{2}{N}(k+1)} g(x)dx + 2N \sum_{k=0}^{\frac{N}{2}-1} (-1)^k \int_{\frac{2}{N}k}^{\frac{2}{N}(k+1)} \int_{\frac{2}{N}k}^{y} g(x)dxdy$$

$$= 2 \sum_{k=0}^{\frac{N}{2}-1} (-1)^{k+1} \int_{\frac{2}{N}k}^{\frac{2}{N}(k+1)} g(x)dx + 2N \sum_{k=0}^{\frac{N}{2}-1} (-1)^k \int_{\frac{2}{N}k}^{\frac{2}{N}(k+1)} \int_{\frac{2}{N}k}^{y} g(x)dxdy$$

$$= 2N \sum_{k=0}^{\frac{N}{2}-1} (-1)^k \int_{\frac{2}{N}k}^{\frac{2}{N}(k+1)} \int_{x}^{\frac{2}{N}(k+1)} g(x)dydx + 2 \sum_{k=0}^{\frac{N}{2}-1} (-1)^{k+1} \int_{\frac{2}{N}k}^{\frac{2}{N}(k+1)} g(x)dx$$

$$= 2N \sum_{k=0}^{\frac{N}{2}-1} (-1)^k \int_{\frac{2}{N}k}^{\frac{2}{N}(k+1)} (\frac{2}{N}(k+1) - x)g(x)dx - 2N \sum_{k=0}^{\frac{N}{2}-1} (-1)^k \int_{\frac{2}{N}k}^{\frac{2}{N}(k+1)} \frac{1}{N}g(x)dx$$

$$= 2N \sum_{k=0}^{\frac{N}{2}-1} (-1)^k \int_{\frac{2}{N}k}^{\frac{2}{N}(k+1)} (\frac{2k+1}{N} - x)g(x)dx = 2N \sum_{k=0}^{\frac{N}{2}-1} (-1)^k \int_{-\frac{1}{N}}^{\frac{1}{N}} xg(\frac{2k+1}{N} - x)dx$$

$$= 2N \sum_{k=0}^{\frac{N}{2}-1} (-1)^k \int_{0}^{\frac{1}{N}} x\left[g(\frac{2k+1}{N} - x) - g(\frac{2k+1}{N} + x)\right]dx$$

Setting this into (10.2), we get

$$\lambda_t^{(N)} = \int_{-1}^{1} g(y)dy \left(1 + \cos\left[\frac{2\pi t}{N}\right]\right) + \left(c_0^{(N)} + (-1)^t c_{\frac{N}{2}}^{(N)} + 2 \sum_{k=1}^{\frac{N}{2}-1} c_k^{(N)} \cos\left[\frac{2\pi kt}{N}\right]\right)$$

$$+ 2N \sum_{k=0}^{\frac{N}{2}-1} (-1)^k \int_{0}^{\frac{1}{N}} x\left[g(\frac{2k+1}{N} - x) - g(\frac{2k+1}{N} + x)\right]dx \left(1 - \cos\left[\frac{2\pi t}{N}\right]\right)$$

$$(10.5)$$

10.2 The Fourier Sum

The hardest work is to estimate the second part, given by

$$c_0^{(N)} + (-1)^t c_{\frac{N}{2}}^{(N)} + 2 \sum_{k=1}^{\frac{N}{2}-1} c_k^{(N)} \cos\left[\frac{2\pi kt}{N}\right] = \sum_{k=0}^{\frac{N}{2}-1} c_k^{(N)} \cos\left[\frac{2\pi kt}{N}\right] + \sum_{k=1}^{\frac{N}{2}} c_k^{(N)} \cos\left[\frac{2\pi kt}{N}\right]$$

$$= \sum_{k=0}^{\frac{N}{2}-1} \left(c_k^{(N)} \cos\left[\frac{2\pi kt}{N}\right] + c_{k+1}^{(N)} \cos\left[\frac{2\pi (k+1)t}{N}\right]\right) = T_{t,N}^{(1)} + T_{t,N}^{(2)}, \quad (10.6)$$

where

$$T_{t,N}^{(1)} = \sum_{k=0}^{\frac{N}{2}-1} c_k^{(N)} \left(\cos\left[\frac{2\pi kt}{N}\right] + \cos\left[\frac{2\pi (k+1)t}{N}\right] \right) \tag{10.7}$$

$$T_{t,N}^{(2)} = \sum_{k=0}^{\frac{N}{2}-1} (c_{k+1}^{(N)} - c_k^{(N)}) \cos\left[\frac{2\pi (k+1)t}{N}\right] \tag{10.8}$$

for every even $N \in \mathbb{N}$, $t \in \{1, \ldots, \frac{N}{2} - 1\}$. For given N, t, we define

$$\delta = \delta_{t,N} = 1 - \frac{2t}{N} > 0. \tag{10.9}$$

Then we have

Lemma 10.1 *Let $N \in \mathbb{N}$ be a multiple of 4 (i.e., $\frac{N}{2}$ is even), $t \in \{1, \ldots, \frac{N}{2} - 1\}$, δ as given by (10.9). Then*

$$|T_{t,N}^{(1)}| \le C(\delta^3 + \frac{\delta^2}{N} + \frac{\delta}{N^2}).$$

Before we prove this, we formulate two simple estimates that will be needed throughout this chapter:

Lemma 10.2 *For every even $N \in \mathbb{N}$, $k = 0, \ldots, N - 1$, we have*

$$|c_k^{(N)}| \le \frac{C}{N}, \tag{10.10}$$

and for $k = 0, \ldots, N - 2$, we have

$$|c_k^{(N)} - c_{k+1}^{(N)}| \le \frac{C}{N^2}, \tag{10.11}$$

where C does not depend on N, k.

Proof: Using (6.12) and the boundedness of g, we deduce

$$|c_k^{(N)}| = \left| N \int_0^{\frac{2}{N}} \int_{\frac{2}{N}k}^{\frac{2}{N}(k+1)} g(x-y)dxdy \right| \le N\frac{4}{N^2}C \le \frac{C}{N}.$$

Furthermore, using the Lipschitz continuity of g, we obtain

$$|c_k^{(N)} - c_{k+1}^{(N)}| = N\left| \int_0^{\frac{2}{N}} \int_{\frac{2}{N}k}^{\frac{2}{N}(k+1)} g(x-y)dxdy - \int_0^{\frac{2}{N}} \int_{\frac{2}{N}(k+1)}^{\frac{2}{N}(k+2)} g(x-y)dxdy \right|$$

$$= N\left| \int_0^{\frac{2}{N}} \int_{\frac{2}{N}k}^{\frac{2}{N}(k+1)} (g(x-y) - g(x-y+\frac{2}{N}))dxdy \right|$$

$$\leq N \int_0^{\frac{2}{N}} \int_{\frac{2}{N}k}^{\frac{2}{N}(k+1)} |g(x-y) - g(x-y+\frac{2}{N})| dx dy$$

$$\leq CN \frac{4}{N^2} \cdot \frac{2}{N} \leq \frac{C}{N^2}.$$

$\square$

Proof of Lemma 10.1: For $k = 0, \ldots, \frac{N}{2} - 1$, we have

$$\cos\left[\frac{2\pi kt}{N}\right] = \cos(\pi k(1 - \delta)) = (-1)^k \cos(k\pi\delta)$$

$$\cos\left[\frac{2\pi(k+1)t}{N}\right] = (-1)^{k+1} \cos((k+1)\pi\delta),$$

so that

$$\cos\left[\frac{2\pi kt}{N}\right] + \cos\left[\frac{2\pi(k+1)t}{N}\right] = (-1)^{k+1}(\cos((k+1)\pi\delta) - \cos(k\pi\delta))$$

$$= (-1)^k \pi\delta \sin(k\pi\delta) + \frac{1}{2}(-1)^k \pi^2\delta^2 \cos(\xi_{k,N})$$

for some $\xi_{k,N} \in [k\pi\delta, (k+1)\pi\delta]$. Using the fact that $\frac{N}{2}$ is even, we obtain

$$T_{t,N}^{(1)} = \pi\delta \sum_{k=0}^{\frac{N}{2}-1} (-1)^k c_k^{(N)} \sin(k\pi\delta) + \frac{1}{2}\pi^2\delta^2 \sum_{k=0}^{\frac{N}{2}-1} (-1)^k c_k^{(N)} \cos(\xi_{k,N})$$

$$= \pi\delta \sum_{\substack{k=0 \\ k \text{ even}}}^{\frac{N}{2}-1} (c_k \sin(k\pi\delta) - c_{k+1} \sin((k+1)\pi\delta)) + \frac{1}{2}\pi^2\delta^2 \sum_{k=0}^{\frac{N}{2}-1} (-1)^k c_k^{(N)} \cos(\xi_{k,N})$$

$$= \pi\delta \sum_{\substack{k=0 \\ k \text{ even}}}^{\frac{N}{2}-1} (c_k - c_{k+1}) \sin(k\pi\delta) + \pi\delta \sum_{\substack{k=0 \\ k \text{ even}}}^{\frac{N}{2}-1} c_{k+1}(\sin(k\pi\delta) - \sin((k+1)\pi\delta))$$

$$+ \frac{1}{2}\pi^2\delta^2 \sum_{k=0}^{\frac{N}{2}-1} (-1)^k c_k^{(N)} \cos(\xi_{k,N}), \tag{10.12}$$

where, by definition of $c_k^{(N)}$ (cf. (6.12)) and due to $\frac{N}{2}$ being even,

$$\frac{1}{2}\pi^2\delta^2 \sum_{k=0}^{\frac{N}{2}-1} (-1)^k c_k^{(N)} \cos(\xi_{k,N}) = \frac{1}{2}\pi^2\delta^2 \sum_{\substack{k=0 \\ k \text{ even}}}^{\frac{N}{2}-1} (c_k^{(N)} \cos(\xi_{k,N}) - c_{k+1}^{(N)} \cos(\xi_{k+1,N}))$$

$$= \frac{1}{2}\pi^2\delta^2 \sum_{\substack{k=0 \\ k \text{ even}}}^{\frac{N}{2}-1} (c_k^{(N)} - c_{k+1}^{(N)}) \cos(\xi_{k,N}) + \frac{1}{2}\pi^2\delta^2 \sum_{\substack{k=0 \\ k \text{ even}}}^{\frac{N}{2}-1} c_{k+1}^{(N)} (\cos(\xi_{k,N}) - \cos(\xi_{k+1,N})).$$

Since $\xi_{k,N} \in [k\pi\delta, (k+1)\pi\delta]$, $\xi_{k+1,N} \in [(k+1)\pi\delta, (k+2)\pi\delta]$, we have

$$|\xi_{k,N} - \xi_{k+1,N}| \leq C\delta,$$

and using (10.10) and (10.11), we obtain

$$\left| \frac{1}{2}\pi^2\delta^2 \sum_{\substack{k=0 \\ k\text{ even}}}^{\frac{N}{2}-1} (-1)^k c_k^{(N)} \cos(\xi_{k,N}) \right| \leq \frac{\pi^2\delta^2}{2} \sum_{\substack{k=0 \\ k\text{ even}}}^{\frac{N}{2}-1} \frac{C}{N^2} + \frac{\pi^2\delta^2}{2} \sum_{\substack{k=0 \\ k\text{ even}}}^{\frac{N}{2}-1} \frac{C}{N}|\cos(\xi_{k,N}) - \cos(\xi_{k+1,N})|$$

$$\leq C\frac{\delta^2}{N} + C\frac{\delta^2}{N} \sum_{\substack{k=0 \\ k\text{ even}}}^{\frac{N}{2}-1} |\xi_{k,N} - \xi_{k+1,N}| \leq C\frac{\delta^2}{N} + C\frac{\delta^2}{N} \sum_{\substack{k=0 \\ k\text{ even}}}^{\frac{N}{2}-1} C\delta \leq C\frac{\delta^2}{N} + C\delta^3.$$

Setting this into (10.12), we get

$$T_{t,N}^{(1)} = \pi\delta \sum_{\substack{k=2 \\ k\text{ even}}}^{\frac{N}{2}-2} (c_k^{(N)} - c_{k+1}^{(N)}) \sin(k\pi\delta)$$

$$+ \pi\delta \sum_{\substack{k=0 \\ k\text{ even}}}^{\frac{N}{2}-1} c_{k+1}^{(N)}(\sin(k\pi\delta) - \sin((k+1)\pi\delta)) + O(\frac{\delta^2}{N}) + O(\delta^3), \qquad (10.13)$$

with the new limits in the first sum resulting from $\sin(0) = 0$ and from the fact that $\frac{N}{2} - 1$ is odd. For $k \in \{2, \ldots, \frac{N}{2} - 2\}$, consider

$$c_k^{(N)} - c_{k+1}^{(N)} = -N \int_0^{\frac{2}{N}} \int_{\frac{2}{N}k}^{\frac{2}{N}(k+1)} g(x-y)dxdy + N \int_0^{\frac{2}{N}} \int_{\frac{2}{N}(k+1)}^{\frac{2}{N}(k+2)} g(x-y)dxdy$$

$$= N \int_0^{\frac{2}{N}} \int_{\frac{2}{N}k}^{\frac{2}{N}(k+1)} \left[g(x-y+\frac{2}{N}) - g(x-y) \right] dxdy.$$

Since

$$0 < \frac{2}{N}(k-1) \leq x - y \leq x - y + \frac{2}{N} \leq \frac{2}{N}(k+2) \leq 1,$$

we can make use of the smoothness of g on $[0,1]$, and for every x, y we can write the integrand as

$$g(x-y+\frac{2}{N}) - g(x-y) = \frac{2}{N}g'(x-y) + \frac{2}{N^2}g''(\rho_{k,N}(x,y)),$$

where $\rho_{k,N}(x,y) \in [x-y, x-y+\frac{2}{N}] \subset [\frac{2}{N}(k-1), \frac{2}{N}(k+2)]$, so that

$$c_k^{(N)} - c_{k+1}^{(N)} = 2 \int_0^{\frac{2}{N}} \int_{\frac{2}{N}k}^{\frac{2}{N}(k+1)} g'(x-y)dxdy + \frac{2}{N} \int_0^{\frac{2}{N}} \int_{\frac{2}{N}k}^{\frac{2}{N}(k+1)} g''(\rho_{k,N}(x,y))dxdy$$

$$= 2 \int_0^{\frac{2}{N}} \left[g(\frac{2}{N}(k+1) - y) - g(\frac{2}{N}k - y) \right] dy + \frac{2}{N} \int_0^{\frac{2}{N}} \int_{\frac{2}{N}k}^{\frac{2}{N}(k+1)} g''(\rho_{k,N}(x,y))dxdy$$

$$= 2 \int_0^{\frac{2}{N}} \left[g(\frac{2}{N}(k+1) - y) - g(\frac{2}{N}k - y) \right] dy + O(\frac{1}{N^3})$$

due to the boundedness of g''. Set into (10.13), this yields

$$T_{t,N}^{(1)} = 2\pi\delta \sum_{\substack{k=2 \\ k \text{ even}}}^{\frac{N}{2}-2} \int_0^{\frac{2}{N}} \left[g(\frac{2}{N}(k+1) - y) - g(\frac{2}{N}k - y) \right] dy \sin(k\pi\delta) + O(\frac{\delta}{N^3}) \sum_{\substack{k=2 \\ k \text{ even}}}^{\frac{N}{2}-2} \sin(k\pi\delta)$$

$$+ \pi\delta \sum_{\substack{k=0 \\ k \text{ even}}}^{\frac{N}{2}-1} c_{k+1}^{(N)}(\sin(k\pi\delta) - \sin((k+1)\pi\delta)) + O(\frac{\delta^2}{N}) + O(\delta^3)$$

$$= 2\pi\delta \sum_{\substack{k=2 \\ k \text{ even}}}^{\frac{N}{2}-2} \int_0^{\frac{2}{N}} \left[g(\frac{2}{N}(k+1) - y) - g(\frac{2}{N}k - y) \right] dy \sin(k\pi\delta)$$

$$+ \pi\delta \sum_{\substack{k=0 \\ k \text{ even}}}^{\frac{N}{2}-1} c_{k+1}^{(N)}(\sin(k\pi\delta) - \sin((k+1)\pi\delta)) + O(\frac{\delta}{N^2}) + O(\frac{\delta^2}{N}) + O(\delta^3). \qquad (10.14)$$

To modify the first term, we note that $\frac{2}{N}(k+1) - y$, $\frac{2}{N}k - y \in (0,1)$ for every $k \in \{2, \ldots, \frac{N}{2} - 2\}$ and $y \in [0, \frac{2}{N}]$, so that due to the smoothness of g we can write the integrands as

$$g(\frac{2}{N}(k+1) - y) - g(\frac{2}{N}k - y) = \frac{2}{N}g'(\frac{2}{N}k - y) + \frac{2}{N^2}g''(\zeta_{k,N}(y) - y)$$

with $\zeta_{k,N}(y) \in [\frac{2}{N}k, \frac{2}{N}(k+1)]$ for every $k \in \{2, \ldots, \frac{N}{2} - 2\}$ and $y \in [0, \frac{2}{N}]$. Hence

$$2\pi\delta \sum_{\substack{k=2 \\ k \text{ even}}}^{\frac{N}{2}-2} \int_0^{\frac{2}{N}} \left[g(\frac{2}{N}(k+1) - y) - g(\frac{2}{N}k - y) \right] dy \sin(k\pi\delta)$$

$$= 4\pi\frac{\delta}{N} \sum_{\substack{k=2 \\ k \text{ even}}}^{\frac{N}{2}-2} \int_0^{\frac{2}{N}} g'(\frac{2}{N}k - y) dy \sin(k\pi\delta) + 4\pi\frac{\delta}{N^2} \sum_{\substack{k=2 \\ k \text{ even}}}^{\frac{N}{2}-2} \int_0^{\frac{2}{N}} g''(\zeta_{k,N}(y) - y) dy \sin(k\pi\delta)$$

$$= 4\pi\frac{\delta}{N} \sum_{\substack{k=2 \\ k \text{ even}}}^{\frac{N}{2}-2} \int_{\frac{2}{N}(k-1)}^{\frac{2}{N}k} g'(y) dy \sin(k\pi\delta) + O(\frac{\delta}{N^2}), \qquad (10.15)$$

since the boundedness of g'' implies

$$\left| 4\pi\frac{\delta}{N^2} \sum_{\substack{k=2 \\ k \text{ even}}}^{\frac{N}{2}-2} \int_0^{\frac{2}{N}} g''(\zeta_{k,N}(y) - y) dy \sin(k\pi\delta) \right| \leq C\frac{\delta}{N^2} \sum_{\substack{k=2 \\ k \text{ even}}}^{\frac{N}{2}-2} \frac{2}{N} \leq C\frac{\delta}{N^2}.$$

The first term can be remodelled as

$$4\pi \frac{\delta}{N} \sum_{\substack{k=2 \\ k \, even}}^{\frac{N}{2}-2} \int_{\frac{2}{N}(k-1)}^{\frac{2}{N}k} g'(y)dy \sin(k\pi\delta) = 4\pi \frac{\delta}{N} \sum_{\substack{k=2 \\ k \, even}}^{\frac{N}{2}-2} \int_{\frac{2}{N}(k-1)}^{\frac{2}{N}k} g'(y) \sin(\frac{N}{2}\pi\delta y)dy$$

$$+ 4\pi \frac{\delta}{N} \sum_{\substack{k=2 \\ k \, even}}^{\frac{N}{2}-2} \int_{\frac{2}{N}(k-1)}^{\frac{2}{N}k} g'(y)(\sin(k\pi\delta) - \sin(\frac{N}{2}\pi\delta y))dy$$

$$= 4\pi \frac{\delta}{N} \sum_{\substack{k=2 \\ k \, even}}^{\frac{N}{2}-2} \int_{\frac{2}{N}(k-1)}^{\frac{2}{N}k} g'(y) \sin(\frac{N}{2}\pi\delta y)dy + O(\frac{\delta^2}{N}), \qquad (10.16)$$

since for $y \in [\frac{2}{N}(k-1), \frac{2}{N}k]$ $(k \in \{2, \ldots, \frac{N}{2} - 2\})$, we have

$$|\sin(k\pi\delta) - \sin(\frac{N}{2}\pi\delta y)| \leq \pi\delta|k - \frac{N}{2}y| \leq \pi\delta(k - (k-1)) = \pi\delta$$

and thus, due to the boundedness of g',

$$\left| 4\pi \frac{\delta}{N} \sum_{\substack{k=2 \\ k \, even}}^{\frac{N}{2}-2} \int_{\frac{2}{N}(k-1)}^{\frac{2}{N}k} g'(y)(\sin(k\pi\delta) - \sin(\frac{N}{2}\pi\delta y))dy \right| \leq C \frac{\delta}{N} \sum_{\substack{k=2 \\ k \, even}}^{\frac{N}{2}-2} \frac{2}{N} \cdot \delta \leq C \frac{\delta^2}{N}.$$

Setting (10.16) into (10.14), we obtain

$$T_{t,N}^{(1)} = 4\pi \frac{\delta}{N} \sum_{\substack{k=2 \\ k \, even}}^{\frac{N}{2}-2} \int_{\frac{2}{N}(k-1)}^{\frac{2}{N}k} g'(y) \sin(\frac{N}{2}\pi\delta y)dy$$

$$+ \pi\delta \sum_{\substack{k=0 \\ k \, even}}^{\frac{N}{2}-1} c_{k+1}^{(N)}(\sin(k\pi\delta) - \sin((k+1)\pi\delta)) + O(\frac{\delta}{N^2}) + O(\frac{\delta^2}{N}) + O(\delta^3). \quad (10.17)$$

Recalling that $\frac{N}{2}$ is even, we will examine the first sum further:

$$4\pi \frac{\delta}{N} \sum_{\substack{k=2 \\ k \, even}}^{\frac{N}{2}-2} \int_{\frac{2}{N}(k-1)}^{\frac{2}{N}k} g'(y) \sin(\frac{N}{2}\pi\delta y)dy = 4\pi \frac{\delta}{N} \sum_{\substack{k=1 \\ k \, odd}}^{\frac{N}{2}-3} \int_{\frac{2}{N}k}^{\frac{2}{N}(k+1)} g'(y) \sin(\frac{N}{2}\pi\delta y)dy$$

$$= 2\pi \frac{\delta}{N} \sum_{\substack{k=1 \\ k \, odd}}^{\frac{N}{2}-3} \int_{\frac{2}{N}k}^{\frac{2}{N}(k+1)} g'(y) \sin(\frac{N}{2}\pi\delta y)dy + 2\pi \frac{\delta}{N} \sum_{k=0}^{\frac{N}{2}-1} \int_{\frac{2}{N}k}^{\frac{2}{N}(k+1)} g'(y) \sin(\frac{N}{2}\pi\delta y)dy$$

$$- 2\pi \frac{\delta}{N} \int_{1-\frac{2}{N}}^{1} g'(y) \sin(\frac{N}{2}\pi\delta y)dy - 2\pi \frac{\delta}{N} \sum_{\substack{k=0 \\ k \, even}}^{\frac{N}{2}-1} \int_{\frac{2}{N}k}^{\frac{2}{N}(k+1)} g'(y) \sin(\frac{N}{2}\pi\delta y)dy$$

$$= 2\pi\frac{\delta}{N}\int_0^1 g'(y)\sin(\frac{N}{2}\pi\delta y)dy + 2\pi\frac{\delta}{N}\sum_{\substack{k=1\\k\ odd}}^{\frac{N}{2}-1}\int_{\frac{2}{N}k}^{\frac{2}{N}(k+1)} g'(y)\sin(\frac{N}{2}\pi\delta y)dy$$

$$- 4\pi\frac{\delta}{N}\int_{1-\frac{2}{N}}^1 g'(y)\sin(\frac{N}{2}\pi\delta y)dy - 2\pi\frac{\delta}{N}\sum_{\substack{k=0\\k\ even}}^{\frac{N}{2}-1}\int_{\frac{2}{N}k}^{\frac{2}{N}(k+1)} g'(y)\sin(\frac{N}{2}\pi\delta y)dy$$

$$= 2\pi\frac{\delta}{N}\int_0^1 g'(y)\sin(\frac{N}{2}\pi\delta y)dy - 4\pi\frac{\delta}{N}\int_{1-\frac{2}{N}}^1 g'(y)\sin(\frac{N}{2}\pi\delta y)dy$$

$$+ 2\pi\frac{\delta}{N}\sum_{\substack{k=1\\k\ odd}}^{\frac{N}{2}-1}\left[\int_{\frac{2}{N}k}^{\frac{2}{N}(k+1)} g'(y)\sin(\frac{N}{2}\pi\delta y)dy - \int_{\frac{2}{N}(k-1)}^{\frac{2}{N}k} g'(y)\sin(\frac{N}{2}\pi\delta y)dy\right]$$

$$= 2\pi\frac{\delta}{N}\int_0^1 g'(y)\sin(\frac{N}{2}\pi\delta y)dy + O(\frac{\delta}{N^2}) + O(\frac{\delta^2}{N}), \tag{10.18}$$

since on the one hand, the boundedness of g' yields

$$\left|4\pi\frac{\delta}{N}\int_{1-\frac{2}{N}}^1 g'(y)\sin(\frac{N}{2}\pi\delta y)dy\right| \le C\frac{\delta}{N}\cdot\frac{1}{N} \le C\frac{\delta}{N^2},$$

and on the other hand, we get, using the Lipschitz continuity and the boundedness of g' on $[0,1]$ (recall that $g \in C^2[0,1]$)

$$\left|\int_{\frac{2}{N}k}^{\frac{2}{N}(k+1)} g'(y)\sin(\frac{N}{2}\pi\delta y)dy - \int_{\frac{2}{N}(k-1)}^{\frac{2}{N}k} g'(y)\sin(\frac{N}{2}\pi\delta y)dy\right|$$

$$= \left|\int_{\frac{2}{N}(k-1)}^{\frac{2}{N}k}\left[g'(y+\frac{2}{N})\sin(\frac{N}{2}\pi\delta(y+\frac{2}{N})) - g'(y)\sin(\frac{N}{2}\pi\delta y)\right]dy\right|$$

$$\le \int_{\frac{2}{N}(k-1)}^{\frac{2}{N}k}\left|g'(y+\frac{2}{N}) - g'(y)\right|\left|\sin(\frac{N}{2}\pi\delta(y+\frac{2}{N}))\right|dy$$

$$+ \int_{\frac{2}{N}(k-1)}^{\frac{2}{N}k}|g'(y)|\left|\sin(\frac{N}{2}\pi\delta(y+\frac{2}{N})) - \sin(\frac{N}{2}\pi\delta y)\right|dy$$

$$\le \frac{2}{N}\cdot\frac{C}{N} + \frac{C}{N}\cdot\pi\delta \le \frac{C}{N^2} + C\frac{\delta}{N}$$

for $k \in \{1,\ldots,\frac{N}{2}-1\}$ odd, which implies

$$\left|2\pi\frac{\delta}{N}\sum_{\substack{k=1\\k\ odd}}^{\frac{N}{2}-1}\left[\int_{\frac{2}{N}k}^{\frac{2}{N}(k+1)} g'(y)\sin(\frac{N}{2}\pi\delta y)dy - \int_{\frac{2}{N}(k-1)}^{\frac{2}{N}k} g'(y)\sin(\frac{N}{2}\pi\delta y)dy\right]\right|$$

$$\le C\frac{\delta}{N}\sum_{\substack{k=1\\k\ odd}}^{\frac{N}{2}-1}(\frac{C}{N^2} + C\frac{\delta}{N}) \le C\frac{\delta}{N^2} + C\frac{\delta^2}{N},$$

so the last step in (10.18) holds true. Combining latter with (10.17), we obtain

$$T_{t,N}^{(1)} = 2\pi \frac{\delta}{N} \int_0^1 g'(y) \sin(\frac{N}{2}\pi\delta y) dy$$

$$+ \pi\delta \sum_{\substack{k=0 \\ k \text{ even}}}^{\frac{N}{2}-1} c_{k+1}^{(N)}(\sin(k\pi\delta) - \sin((k+1)\pi\delta)) + O(\frac{\delta}{N^2}) + O(\frac{\delta^2}{N}) + O(\delta^3). \quad (10.19)$$

To estimate the sum, we observe that for every k,

$$\sin(k\pi\delta) - \sin((k+1)\pi\delta) = -\pi\delta\cos(k\pi\delta) + \frac{1}{2}\pi^2\delta^2\sin(\eta_{N,k})$$

for some $\eta_{N,k} \in [k\pi\delta, (k+1)\pi\delta]$, so that

$$\pi\delta \sum_{\substack{k=0 \\ k \text{ even}}}^{\frac{N}{2}-1} c_{k+1}^{(N)}(\sin(k\pi\delta) - \sin((k+1)\pi\delta)) =$$

$$= -\pi^2\delta^2 \sum_{\substack{k=0 \\ k \text{ even}}}^{\frac{N}{2}-1} c_{k+1}^{(N)}\cos(k\pi\delta) + \frac{\pi^3}{2}\delta^3 \sum_{\substack{k=0 \\ k \text{ even}}}^{\frac{N}{2}-1} c_{k+1}^{(N)}\sin(\eta_{N,k})$$

$$= -\pi^2\delta^2 \sum_{\substack{k=0 \\ k \text{ even}}}^{\frac{N}{2}-1} c_{k+1}^{(N)}\cos(k\pi\delta) + O(\delta^3),$$

since by (10.10)

$$\left| \frac{\pi^3}{2}\delta^3 \sum_{\substack{k=0 \\ k \text{ even}}}^{\frac{N}{2}-1} c_{k+1}^{(N)}\sin(\eta_{N,k}) \right| \leq C\delta^3 \sum_{\substack{k=0 \\ k \text{ even}}}^{\frac{N}{2}-1} \frac{C}{N} \leq C\delta^3.$$

(10.19) now takes the form

$$T_{t,N}^{(1)} = 2\pi \frac{\delta}{N} \int_0^1 g'(y) \sin(\frac{N}{2}\pi\delta y) dy - \pi^2\delta^2 \sum_{\substack{k=0 \\ k \text{ even}}}^{\frac{N}{2}-1} c_{k+1}^{(N)}\cos(k\pi\delta)$$

$$+ O(\frac{\delta}{N^2}) + O(\frac{\delta^2}{N}) + O(\delta^3). \quad (10.20)$$

Now consider, using (6.12),

$$-\pi^2\delta^2 \sum_{\substack{k=0 \\ k \text{ even}}}^{\frac{N}{2}-1} c_{k+1}^{(N)}\cos(k\pi\delta) = \pi^2\delta^2 N \sum_{\substack{k=0 \\ k \text{ even}}}^{\frac{N}{2}-1} \int_0^{\frac{2}{N}} \int_{\frac{2}{N}(k+1)}^{\frac{2}{N}(k+2)} g(x-y) dx dy \cos(k\pi\delta)$$

$$= \pi^2\delta^2 N \sum_{\substack{k=0 \\ k \text{ even}}}^{\frac{N}{2}-1} \int_0^{\frac{2}{N}} \int_{\frac{2}{N}(k+1)}^{\frac{2}{N}(k+2)} (g(x-y) - g(x)) dx dy \cos(k\pi\delta)$$

$$+ 2\pi^2\delta^2 \sum_{\substack{k=0 \\ k \text{ even}}}^{\frac{N}{2}-1} \int_{\frac{2}{N}(k+1)}^{\frac{2}{N}(k+2)} g(x)dx \cos(k\pi\delta)$$

$$= 2\pi^2\delta^2 \sum_{\substack{k=0 \\ k \text{ even}}}^{\frac{N}{2}-1} \int_{\frac{2}{N}(k+1)}^{\frac{2}{N}(k+2)} g(x)dx \cos(k\pi\delta) + O(\frac{\delta^2}{N}), \qquad (10.21)$$

because the Lipschitz continuity of g implies

$$\left| \pi^2\delta^2 N \sum_{\substack{k=0 \\ k \text{ even}}}^{\frac{N}{2}-1} \int_0^{\frac{2}{N}} \int_{\frac{2}{N}(k+1)}^{\frac{2}{N}(k+2)} (g(x-y) - g(x))dxdy \cos(k\pi\delta) \right|$$

$$\leq C\delta^2 N \sum_{\substack{k=0 \\ k \text{ even}}}^{\frac{N}{2}-1} \int_0^{\frac{2}{N}} \int_{\frac{2}{N}(k+1)}^{\frac{2}{N}(k+2)} |y|dxdy$$

$$\leq C\delta^2 N \sum_{\substack{k=0 \\ k \text{ even}}}^{\frac{N}{2}-1} \int_0^{\frac{2}{N}} \int_{\frac{2}{N}(k+1)}^{\frac{2}{N}(k+2)} \frac{2}{N}dxdy \leq C\delta^2 N \sum_{\substack{k=0 \\ k \text{ even}}}^{\frac{N}{2}-1} \frac{8}{N^3} \leq C\frac{\delta^2}{N}.$$

We set (10.21) into (10.20) and obtain

$$T_{t,N}^{(1)} = 2\pi\frac{\delta}{N} \int_0^1 g'(y) \sin(\frac{N}{2}\pi\delta y)dy + 2\pi^2\delta^2 \sum_{\substack{k=0 \\ k \text{ even}}}^{\frac{N}{2}-1} \int_{\frac{2}{N}(k+1)}^{\frac{2}{N}(k+2)} g(x)dx \cos(k\pi\delta)$$

$$+ O(\frac{\delta}{N^2}) + O(\frac{\delta^2}{N}) + O(\delta^3). \qquad (10.22)$$

Similar as above, we approximate the sum by a "Fourier-type" integral as follows (using that $\frac{N}{2} - 1$ is odd):

$$2\pi^2\delta^2 \sum_{\substack{k=0 \\ k \text{ even}}}^{\frac{N}{2}-1} \int_{\frac{2}{N}(k+1)}^{\frac{2}{N}(k+2)} g(x)dx \cos(k\pi\delta) = 2\pi^2\delta^2 \sum_{\substack{k=0 \\ k \text{ even}}}^{\frac{N}{2}-1} \int_{\frac{2}{N}(k+1)}^{\frac{2}{N}(k+2)} g(x) \cos(\frac{N}{2}\pi\delta x)dx$$

$$+ 2\pi^2\delta^2 \sum_{\substack{k=0 \\ k \text{ even}}}^{\frac{N}{2}-1} \int_{\frac{2}{N}(k+1)}^{\frac{2}{N}(k+2)} g(x)(\cos(k\pi\delta) - \cos(\frac{N}{2}\pi\delta x))dx$$

$$= 2\pi^2\delta^2 \sum_{\substack{k=1 \\ k \text{ odd}}}^{\frac{N}{2}-1} \int_{\frac{2}{N}k}^{\frac{2}{N}(k+1)} g(x) \cos(\frac{N}{2}\pi\delta x)dx + O(\delta^3), \qquad (10.23)$$

since for $k = 0, \ldots, \frac{N}{2} - 1$, $x \in [\frac{2}{N}(k+1), \frac{2}{N}(k+2)]$, we have

$$|\cos(k\pi\delta) - \cos(\frac{N}{2}\pi\delta x)| \leq \pi\delta|k - \frac{N}{2}x| \leq \pi\delta((k+2) - k) = C\delta,$$

which together with the fact that g is bounded implies that

$$\left|2\pi^2\delta^2 \sum_{\substack{k=0 \\ k \text{ even}}}^{\frac{N}{2}-1} \int_{\frac{2}{N}(k+1)}^{\frac{2}{N}(k+2)} g(x)(\cos(k\pi\delta) - \cos(\frac{N}{2}\pi\delta x))dx\right| \leq C\delta^3 \sum_{\substack{k=0 \\ k \text{ even}}}^{\frac{N}{2}-1} \frac{2}{N} \leq C\delta^3,$$

and (10.23) holds true. We rewrite the sum as

$$2\pi^2\delta^2 \sum_{\substack{k=1 \\ k \text{ odd}}}^{\frac{N}{2}-1} \int_{\frac{2}{N}k}^{\frac{2}{N}(k+1)} g(x) \cos(\frac{N}{2}\pi\delta x)dx$$

$$= \pi^2\delta^2 \sum_{\substack{k=1 \\ k \text{ odd}}}^{\frac{N}{2}-1} \int_{\frac{2}{N}k}^{\frac{2}{N}(k+1)} g(x) \cos(\frac{N}{2}\pi\delta x)dx + \pi^2\delta^2 \int_0^1 g(x) \cos(\frac{N}{2}\pi\delta x)dx$$

$$- \pi^2\delta^2 \sum_{\substack{k=0 \\ k \text{ even}}}^{\frac{N}{2}-1} \int_{\frac{2}{N}k}^{\frac{2}{N}(k+1)} g(x) \cos(\frac{N}{2}\pi\delta x)dx$$

$$= \pi^2\delta^2 \int_0^1 g(x) \cos(\frac{N}{2}\pi\delta x)dx$$

$$+ \pi^2\delta^2 \sum_{\substack{k=0 \\ k \text{ even}}}^{\frac{N}{2}-1} \left[\int_{\frac{2}{N}(k+1)}^{\frac{2}{N}(k+2)} g(x) \cos(\frac{N}{2}\pi\delta x)dx - \int_{\frac{2}{N}k}^{\frac{2}{N}(k+1)} g(x) \cos(\frac{N}{2}\pi\delta x)dx\right]$$

$$= \pi^2\delta^2 \int_0^1 g(x) \cos(\frac{N}{2}\pi\delta x)dx$$

$$+ \pi^2\delta^2 \sum_{\substack{k=0 \\ k \text{ even}}}^{\frac{N}{2}-1} \int_{\frac{2}{N}k}^{\frac{2}{N}(k+1)} \left[g(x + \frac{2}{N}) \cos(\frac{N}{2}\pi\delta(x + \frac{2}{N})) - g(x) \cos(\frac{N}{2}\pi\delta x)\right]dx$$

$$= \pi^2\delta^2 \int_0^1 g(x) \cos(\frac{N}{2}\pi\delta x)dx + O(\frac{\delta^2}{N}) + O(\delta^3), \tag{10.24}$$

since

$$\left|g(x + \frac{2}{N}) \cos(\frac{N}{2}\pi\delta(x + \frac{2}{N})) - g(x) \cos(\frac{N}{2}\pi\delta x)\right|$$

$$\leq \left|g(x + \frac{2}{N}) - g(x)\right|\left|\cos(\frac{N}{2}\pi\delta(x + \frac{2}{N}))\right|$$

$$+ |g(x)|\left|\cos(\frac{N}{2}\pi\delta(x + \frac{2}{N})) - \cos(\frac{N}{2}\pi\delta x)\right| \leq \frac{C}{N} + C\pi\delta \leq \frac{C}{N} + C\delta$$

due to the Lipschitz continuity and boundedness of g, so that

$$\left|\pi^2\delta^2 \sum_{\substack{k=0 \\ k \text{ even}}}^{\frac{N}{2}-1} \int_{\frac{2}{N}k}^{\frac{2}{N}(k+1)} \left[g(x + \frac{2}{N}) \cos(\frac{N}{2}\pi\delta(x + \frac{2}{N})) - g(x) \cos(\frac{N}{2}\pi\delta x)\right]dx\right|$$

$$\leq C\delta^2 \sum_{\substack{k=0 \\ k \text{ even}}}^{\frac{N}{2}-1} \int_{\frac{2}{N}k}^{\frac{2}{N}(k+1)} (\frac{1}{N} + \delta)dx \leq C\delta^2 \sum_{\substack{k=0 \\ k \text{ even}}}^{\frac{N}{2}-1} (\frac{1}{N^2} + \frac{\delta}{N}) \leq C\frac{\delta^2}{N} + C\delta^3,$$

which confirms (10.24). Combining latter with (10.23), we obtain

$$2\pi^2\delta^2 \sum_{\substack{k=0 \\ k \text{ even}}}^{\frac{N}{2}-1} \int_{\frac{2}{N}(k+1)}^{\frac{2}{N}(k+2)} g(x)dx \cos(k\pi\delta) = \pi^2\delta^2 \int_0^1 g(x)\cos(\frac{N}{2}\pi\delta x)dx + O(\frac{\delta^2}{N}) + O(\delta^3),$$

which together with (10.22) implies

$$T_{t,N}^{(1)} = 2\pi\frac{\delta}{N}\int_0^1 g'(y)\sin(\frac{N}{2}\pi\delta y)dy + \pi^2\delta^2 \int_0^1 g(x)\cos(\frac{N}{2}\pi\delta x)dx$$
$$+ O(\frac{\delta}{N^2}) + O(\frac{\delta^2}{N}) + O(\delta^3). \tag{10.25}$$

Recalling (10.9), by partial integration we get

$$2\pi\frac{\delta}{N}\int_0^1 g'(y)\sin(\frac{N}{2}\pi\delta y)dy = -\pi^2\delta^2 \int_0^1 g(y)\cos(\frac{N}{2}\pi\delta y)dy + 2\pi\frac{\delta}{N}g(y)\sin(\frac{N}{2}\pi\delta y)\Big|_{y=0}^{y=1}$$
$$= -\pi^2\delta^2 \int_0^1 g(y)\cos(\frac{N}{2}\pi\delta y)dy + 2\pi\frac{\delta}{N}g(1)\sin(\frac{N}{2}\pi(1 - \frac{2t}{N}))$$
$$= -\pi^2\delta^2 \int_0^1 g(y)\cos(\frac{N}{2}\pi\delta y)dy + 2\pi\frac{\delta}{N}g(1)\sin((\frac{N}{2} - t)\pi)$$
$$= -\pi^2\delta^2 \int_0^1 g(y)\cos(\frac{N}{2}\pi\delta y)dy$$

since $\frac{N}{2} - t \in \mathbb{N}$, and together with (10.25) this yields

$$T_{t,N}^{(1)} = O(\frac{\delta}{N^2}) + O(\frac{\delta^2}{N}) + O(\delta^3),$$

and the proof of Lemma 10.1 is complete. $\square$

$T_{t,N}^{(2)}$, given by (10.8), can be approximated as follows:

Lemma 10.3 Let $N \in \mathbb{N}$ be a multiple of 4 (i.e., $\frac{N}{2}$ is even), $t \in \{1, \ldots, \frac{N}{2} - 1\}$, δ as given by (10.9). Then

$$T_{t,N}^{(2)} = \frac{4}{3N^2}g'(0) + O(\frac{\delta^2}{N}) + O(\frac{\delta}{N^2}) + O(\frac{1}{N^3})$$

Proof: Recalling (10.9), we have

$$\cos(\frac{2(k+1)t\pi}{N}) = \cos((1 - \frac{2t}{N})(k+1)\pi - (k+1)\pi) = (-1)^{k+1}\cos((k+1)\pi\delta),$$

so that, since $\frac{N}{2}$ is even,

$$T_{t,N}^{(2)} = \sum_{k=0}^{\frac{N}{2}-1}(-1)^{k+1}(c_{k+1}^{(N)} - c_k^{(N)})\cos((k+1)\pi\delta)$$

$$= \sum_{\substack{k=0\\k\ even}}^{\frac{N}{2}-1}[(c_{k+2}^{(N)} - c_{k+1}^{(N)})\cos((k+2)\pi\delta) - (c_{k+1}^{(N)} - c_k^{(N)})\cos((k+1)\pi\delta)]$$

$$= \sum_{\substack{k=0\\k\ even}}^{\frac{N}{2}-1}(c_k^{(N)} + c_{k+2}^{(N)} - 2c_{k+1}^{(N)})\cos((k+1)\pi\delta)$$

$$+ \sum_{\substack{k=0\\k\ even}}^{\frac{N}{2}-1}(c_{k+2}^{(N)} - c_{k+1}^{(N)})(\cos((k+2)\pi\delta) - \cos((k+1)\pi\delta))$$

For every k, we will find $\xi_{k,N} \in [(k+1)\pi\delta, k\pi\delta]$ such that

$$\cos((k+2)\pi\delta) - \cos((k+1)\pi\delta) = -\pi\delta\sin((k+1)\pi\delta) - \frac{1}{2}\pi^2\delta^2\cos(\xi_{k,N}),$$

thus

$$T_{t,N}^{(2)} = \sum_{\substack{k=0\\k\ even}}^{\frac{N}{2}-1}(c_k^{(N)} + c_{k+2}^{(N)} - 2c_{k+1}^{(N)})\cos((k+1)\pi\delta) - \pi\delta\sum_{\substack{k=0\\k\ even}}^{\frac{N}{2}-1}(c_{k+2}^{(N)} - c_{k+1}^{(N)})\sin((k+1)\pi\delta)$$

$$- \frac{1}{2}\pi^2\delta^2\sum_{\substack{k=0\\k\ even}}^{\frac{N}{2}-1}(c_{k+2}^{(N)} - c_{k+1}^{(N)})\cos(\xi_{k,N})$$

$$= \sum_{\substack{k=0\\k\ even}}^{\frac{N}{2}-1}(c_k^{(N)} + c_{k+2}^{(N)} - 2c_{k+1}^{(N)})\cos((k+1)\pi\delta) + \pi\delta\sum_{\substack{k=0\\k\ even}}^{\frac{N}{2}-4}(c_{k+1}^{(N)} - c_{k+2}^{(N)})\sin((k+1)\pi\delta)$$

$$+ O(\frac{\delta}{N^2}) + O(\frac{\delta^2}{N}), \tag{10.26}$$

since by (10.11), we have

$$\left|\frac{1}{2}\pi^2\delta^2\sum_{\substack{k=0\\k\ even}}^{\frac{N}{2}-1}(c_{k+2}^{(N)} - c_{k+1}^{(N)})\cos(\xi_{k,N})\right| \leq \frac{1}{2}\pi^2\delta^2\sum_{\substack{k=0\\k\ even}}^{\frac{N}{2}-1}\frac{C}{N^2} \leq C\frac{\delta^2}{N},$$

and the "missing summand" ($k = \frac{N}{2} - 2$, recall that $\frac{N}{2}$ is even) satisfies

$$|\pi\delta(c_{\frac{N}{2}-1}^{(N)} - c_{\frac{N}{2}}^{(N)})\sin((\frac{N}{2} - 1)\pi\delta)| \leq C\frac{\delta}{N^2}$$

due to (10.11). For $k = 0, \ldots, \frac{N}{2} - 4$ even, we examine

$$c_{k+1}^{(N)} - c_{k+2}^{(N)} = N \left[\int_0^{\frac{2}{N}} \int_{\frac{2}{N}(k+2)}^{\frac{2}{N}(k+3)} g(x-y) dx dy - \int_0^{\frac{2}{N}} \int_{\frac{2}{N}(k+1)}^{\frac{2}{N}(k+2)} g(x-y) dx dy \right]$$

$$= N \int_0^{\frac{2}{N}} \int_{\frac{2}{N}(k+1)}^{\frac{2}{N}(k+2)} \left[g(x-y+\frac{2}{N}) - g(x-y) \right] dx dy.$$

Making sure that $[x-y, x-y+\frac{2}{N}] \subset [0,1]$ for every $x \in [\frac{2}{N}(k+1), \frac{2}{N}(k+2)]$, $y \in [0, \frac{2}{N}]$, $k = 0, \ldots, \frac{N}{2} - 4$ even, we obtain, using the smoothness of g on $[0,1]$,

$$g(x-y+\frac{2}{N}) - g(x-y) = \frac{2}{N}g'(x-y) + \frac{2}{N^2}g''(\rho_{k,N}(x,y))$$

with $\rho_{k,N}(x,y) \in [x-y, x-y+\frac{2}{N}]$ for every x, y. Thus

$$c_{k+1}^{(N)} - c_{k+2}^{(N)} = 2 \int_0^{\frac{2}{N}} \int_{\frac{2}{N}(k+1)}^{\frac{2}{N}(k+2)} g'(x-y) dx dy + \frac{2}{N} \int_0^{\frac{2}{N}} \int_{\frac{2}{N}(k+1)}^{\frac{2}{N}(k+2)} g''(\rho_{k,N}(x,y)) dx dy$$

$$= 2 \int_0^{\frac{2}{N}} \left[g(\frac{2}{N}(k+2) - y) - g(\frac{2}{N}(k+1) - y) \right] dy$$

$$+ \frac{2}{N} \int_0^{\frac{2}{N}} \int_{\frac{2}{N}(k+1)}^{\frac{2}{N}(k+2)} g''(\rho_{k,N}(x,y)) dx dy$$

for $k = 0, \ldots, \frac{N}{2} - 4$ even, which set into (10.26) yields

$$T_{t,N}^{(2)} = \sum_{\substack{k=0 \\ k \, even}}^{\frac{N}{2}-1} (c_k^{(N)} + c_{k+2}^{(N)} - 2c_{k+1}^{(N)}) \cos((k+1)\pi\delta)$$

$$+ 2\pi\delta \sum_{\substack{k=0 \\ k \, even}}^{\frac{N}{2}-4} \int_0^{\frac{2}{N}} \left[g(\frac{2}{N}(k+2) - y) - g(\frac{2}{N}(k+1) - y) \right] dy \sin((k+1)\pi\delta)$$

$$+ 2\pi \frac{\delta}{N} \sum_{\substack{k=0 \\ k \, even}}^{\frac{N}{2}-4} \int_0^{\frac{2}{N}} \int_{\frac{2}{N}(k+1)}^{\frac{2}{N}(k+2)} g''(\rho_{k,N}(x,y)) dx dy \sin((k+1)\pi\delta) + O(\frac{\delta}{N^2}) + O(\frac{\delta^2}{N})$$

$$= 2\pi\delta \sum_{\substack{k=0 \\ k \, even}}^{\frac{N}{2}-1} \int_0^{\frac{2}{N}} \left[g(\frac{2}{N}(k+2) - y) - g(\frac{2}{N}(k+1) - y) \right] dy \sin((k+1)\pi\delta)$$

$$+ \sum_{\substack{k=0 \\ k \, even}}^{\frac{N}{2}-1} (c_k^{(N)} + c_{k+2}^{(N)} - 2c_{k+1}^{(N)}) \cos((k+1)\pi\delta) + O(\frac{\delta}{N^2}) + O(\frac{\delta^2}{N}), \qquad (10.27)$$

since the boundedness of g'' implies

$$\left| 2\pi \frac{\delta}{N} \sum_{\substack{k=0 \\ k \, even}}^{\frac{N}{2}-4} \int_0^{\frac{2}{N}} \int_{\frac{2}{N}(k+1)}^{\frac{2}{N}(k+2)} g''(\rho_{k,N}(x,y)) dx dy \sin((k+1)\pi\delta) \right| \leq C \frac{\delta}{N} \sum_{\substack{k=0 \\ k \, even}}^{\frac{N}{2}-4} \frac{C}{N^2} \leq C \frac{\delta}{N^2},$$

while the "extra summand" $(k = \frac{N}{2} - 2)$ is estimated as follows, using the Lipschitz continuity of g:

$$\left| 2\pi\delta \int_0^{\frac{2}{N}} \left[g(1-y) - g(1 - \frac{2}{N} - y) \right] dy \sin((\frac{N}{2} - 1)\pi\delta) \right| \leq C\delta \int_0^{\frac{2}{N}} \frac{C}{N} dy \leq C\frac{\delta}{N^2}.$$

As for the first sum in (10.27), we check that $[\frac{2}{N}(k+1) - y, \frac{2}{N}(k+2) - y] \subset [0,1]$ for every $y \in [0, \frac{2}{N}]$ and every $k \in \{0, \ldots, \frac{N}{2} - 1\}$ even (recall that $\frac{N}{2} - 1$ is odd), so that for every y, we have

$$g(\frac{2}{N}(k+2) - y) - g(\frac{2}{N}(k+1) - y) = \frac{2}{N}g'(\frac{2}{N}(k+1) - y) + \frac{2}{N^2}g''(\zeta_{k,N}(y))$$

for some $\zeta_{k,N}(y) \in [\frac{2}{N}(k+1) - y, \frac{2}{N}(k+2) - y]$, hence

$$2\pi\delta \sum_{\substack{k=0 \\ k\ even}}^{\frac{N}{2}-1} \int_0^{\frac{2}{N}} \left[g(\frac{2}{N}(k+2) - y) - g(\frac{2}{N}(k+1) - y) \right] dy \sin((k+1)\pi\delta) =$$

$$= 4\pi\frac{\delta}{N} \sum_{\substack{k=0 \\ k\ even}}^{\frac{N}{2}-1} \int_0^{\frac{2}{N}} g'(\frac{2}{N}(k+1) - y) dy \sin((k+1)\pi\delta)$$

$$+ 4\pi\frac{\delta}{N^2} \sum_{\substack{k=0 \\ k\ even}}^{\frac{N}{2}-1} \int_0^{\frac{2}{N}} g''(\zeta_{k,N}(y)) dy \sin((k+1)\pi\delta)$$

$$= 4\pi\frac{\delta}{N} \sum_{\substack{k=0 \\ k\ even}}^{\frac{N}{2}-1} \int_0^{\frac{2}{N}} g'(\frac{2}{N}(k+1) - y) dy \sin((k+1)\pi\delta) + O(\frac{\delta}{N^2}),$$

since g'' is bounded and thus

$$\left| 4\pi\frac{\delta}{N^2} \sum_{\substack{k=0 \\ k\ even}}^{\frac{N}{2}-1} \int_0^{\frac{2}{N}} g''(\zeta_{k,N}(y)) dy \sin((k+1)\pi\delta) \right| \leq C\frac{\delta}{N^2} \sum_{\substack{k=0 \\ k\ even}}^{\frac{N}{2}-1} \frac{C}{N} \leq C\frac{\delta}{N^2}.$$

(10.27) now takes the form

$$T_{t,N}^{(2)} = 4\pi\frac{\delta}{N} \sum_{\substack{k=0 \\ k\ even}}^{\frac{N}{2}-1} \int_0^{\frac{2}{N}} g'(\frac{2}{N}(k+1) - y) dy \sin((k+1)\pi\delta)$$

$$+ \sum_{\substack{k=0 \\ k\ even}}^{\frac{N}{2}-1} (c_k^{(N)} + c_{k+2}^{(N)} - 2c_{k+1}^{(N)}) \cos((k+1)\pi\delta) + O(\frac{\delta}{N^2}) + O(\frac{\delta^2}{N})$$

$$= 4\pi\frac{\delta}{N} \sum_{\substack{k=0 \\ k\ even}}^{\frac{N}{2}-1} \int_{\frac{2}{N}k}^{\frac{2}{N}(k+1)} g'(y) dy \sin((k+1)\pi\delta)$$

$$+ \sum_{\substack{k=0 \\ k \text{ even}}}^{\frac{N}{2}-1} (c_k^{(N)} + c_{k+2}^{(N)} - 2c_{k+1}^{(N)}) \cos((k+1)\pi\delta) + O(\frac{\delta}{N^2}) + O(\frac{\delta^2}{N}). \qquad (10.28)$$

Let us examine the first sum further:

$$4\pi\frac{\delta}{N} \sum_{\substack{k=0 \\ k \text{ even}}}^{\frac{N}{2}-1} \int_{\frac{2}{N}k}^{\frac{2}{N}(k+1)} g'(y)dy \sin((k+1)\pi\delta) = 4\pi\frac{\delta}{N} \sum_{\substack{k=0 \\ k \text{ even}}}^{\frac{N}{2}-1} \int_{\frac{2}{N}k}^{\frac{2}{N}(k+1)} g'(y) \sin(\frac{N}{2}\pi\delta y)dy$$

$$+ 4\pi\frac{\delta}{N} \sum_{\substack{k=0 \\ k \text{ even}}}^{\frac{N}{2}-1} \int_{\frac{2}{N}k}^{\frac{2}{N}(k+1)} g'(y)(\sin((k+1)\pi\delta) - \sin(\frac{N}{2}\pi\delta y))dy$$

$$= 4\pi\frac{\delta}{N} \sum_{\substack{k=0 \\ k \text{ even}}}^{\frac{N}{2}-1} \int_{\frac{2}{N}k}^{\frac{2}{N}(k+1)} g'(y) \sin(\frac{N}{2}\pi\delta y)dy + O(\frac{\delta^2}{N}), \qquad (10.29)$$

since for $k \in \{0, \ldots, \frac{N}{2}-1\}$, $y \in [\frac{2}{N}k, \frac{2}{N}(k+1)]$, we have

$$|\sin((k+1)\pi\delta) - \sin(\frac{N}{2}\pi\delta y)| \leq \pi\delta|k+1 - \frac{N}{2}y| \leq \pi\delta$$

and using the boundedness of g' we obtain

$$\left| 4\pi\frac{\delta}{N} \sum_{\substack{k=0 \\ k \text{ even}}}^{\frac{N}{2}-1} \int_{\frac{2}{N}k}^{\frac{2}{N}(k+1)} g'(y)(\sin((k+1)\pi\delta) - \sin(\frac{N}{2}\pi\delta y))dy \right|$$

$$\leq C\frac{\delta^2}{N} \sum_{\substack{k=0 \\ k \text{ even}}}^{\frac{N}{2}-1} \int_{\frac{2}{N}k}^{\frac{2}{N}(k+1)} |g'(y)|dy \leq C\frac{\delta^2}{N} \sum_{\substack{k=0 \\ k \text{ even}}}^{\frac{N}{2}-1} \frac{2}{N} \leq C\frac{\delta^2}{N}.$$

Furthermore, using the fact that $\frac{N}{2}$ is even,

$$4\pi\frac{\delta}{N} \sum_{\substack{k=0 \\ k \text{ even}}}^{\frac{N}{2}-1} \int_{\frac{2}{N}k}^{\frac{2}{N}(k+1)} g'(y) \sin(\frac{N}{2}\pi\delta y)dy =$$

$$= 2\pi\frac{\delta}{N} \int_0^1 g'(y) \sin(\frac{N}{2}\pi\delta y)dy + 2\pi\frac{\delta}{N} \sum_{\substack{k=0 \\ k \text{ even}}}^{\frac{N}{2}-1} \int_{\frac{2}{N}k}^{\frac{2}{N}(k+1)} g'(y) \sin(\frac{N}{2}\pi\delta y)dy$$

$$- 2\pi\frac{\delta}{N} \sum_{\substack{k=1 \\ k \text{ odd}}}^{\frac{N}{2}-1} \int_{\frac{2}{N}k}^{\frac{2}{N}(k+1)} g'(y) \sin(\frac{N}{2}\pi\delta y)dy$$

$$= 2\pi\frac{\delta}{N} \int_0^1 g'(y) \sin(\frac{N}{2}\pi\delta y)dy$$

$$+ 2\pi \frac{\delta}{N} \sum_{\substack{k=0 \\ k\ even}}^{\frac{N}{2}-2} \left[\int_{\frac{2}{N}k}^{\frac{2}{N}(k+1)} g'(y) \sin(\frac{N}{2}\pi\delta y) dy - \int_{\frac{2}{N}(k+1)}^{\frac{2}{N}(k+2)} g'(y) \sin(\frac{N}{2}\pi\delta y) dy \right]$$

$$= 2\pi \frac{\delta}{N} \int_0^1 g'(y) \sin(\frac{N}{2}\pi\delta y) dy$$

$$+ 2\pi \frac{\delta}{N} \sum_{\substack{k=0 \\ k\ even}}^{\frac{N}{2}-2} \int_{\frac{2}{N}k}^{\frac{2}{N}(k+1)} \left[g'(y) \sin(\frac{N}{2}\pi\delta y) - g'(y+\frac{2}{N}) \sin(\frac{N}{2}\pi\delta(y+\frac{2}{N})) \right] dy$$

$$= 2\pi \frac{\delta}{N} \int_0^1 g'(y) \sin(\frac{N}{2}\pi\delta y) dy + O(\frac{\delta}{N^2}) + O(\frac{\delta^2}{N}), \tag{10.30}$$

since for $k = 0, \ldots, \frac{N}{2}-2$, $y \in [\frac{2}{N}k, \frac{2}{N}(k+1)]$, we have $[y, y+\frac{2}{N}] \subset [\frac{2}{N}k, \frac{2}{N}(k+2)] \subset [0,1]$, so that the Lipschitz continuity and the boundedness of g' on $[0,1]$ yield

$$\left| g'(y) \sin(\frac{N}{2}\pi\delta y) - g'(y+\frac{2}{N}) \sin(\frac{N}{2}\pi\delta(y+\frac{2}{N})) \right|$$

$$\leq |g'(y)| \left| \sin(\frac{N}{2}\pi\delta y) - \sin(\frac{N}{2}\pi\delta(y+\frac{2}{N})) \right| + \left| g'(y) - g'(y+\frac{2}{N}) \right| \left| \sin(\frac{N}{2}\pi\delta(y+\frac{2}{N})) \right|$$

$$\leq C\pi\delta + \frac{C}{N}$$

and thus

$$\left| 2\pi \frac{\delta}{N} \sum_{\substack{k=0 \\ k\ even}}^{\frac{N}{2}-2} \int_{\frac{2}{N}k}^{\frac{2}{N}(k+1)} \left[g'(y) \sin(\frac{N}{2}\pi\delta y) - g'(y+\frac{2}{N}) \sin(\frac{N}{2}\pi\delta(y+\frac{2}{N})) \right] dy \right|$$

$$\leq C\frac{\delta}{N}(C\delta + \frac{C}{N}) \sum_{\substack{k=0 \\ k\ even}}^{\frac{N}{2}-2} \frac{2}{N} \leq C\frac{\delta^2}{N} + C\frac{\delta}{N^2}.$$

Setting (10.30) into (10.29), we obtain

$$4\pi \frac{\delta}{N} \sum_{\substack{k=0 \\ k\ even}}^{\frac{N}{2}-1} \int_{\frac{2}{N}k}^{\frac{2}{N}(k+1)} g'(y) dy \sin((k+1)\pi\delta) = 2\pi \frac{\delta}{N} \int_0^1 g'(y) \sin(\frac{N}{2}\pi\delta y) dy + O(\frac{\delta}{N^2}) + O(\frac{\delta^2}{N}),$$

and from (10.28) we get

$$T_{t,N}^{(2)} = 2\pi \frac{\delta}{N} \int_0^1 g'(y) \sin(\frac{N}{2}\pi\delta y) dy$$

$$+ \sum_{\substack{k=0 \\ k\ even}}^{\frac{N}{2}-1} (c_k^{(N)} + c_{k+2}^{(N)} - 2c_{k+1}^{(N)}) \cos((k+1)\pi\delta) + O(\frac{\delta}{N^2}) + O(\frac{\delta^2}{N}). \tag{10.31}$$

Let $k \in \{2, \ldots, \frac{N}{2} - 4\}$ even, then (cf. (6.12))

$$c_k^{(N)} + c_{k+2}^{(N)} - 2c_{k+1}^{(N)} = -N \int_0^{\frac{2}{N}} \int_{\frac{2}{N}k}^{\frac{2}{N}(k+1)} g(x-y)dxdy$$

$$- N \int_0^{\frac{2}{N}} \int_{\frac{2}{N}(k+2)}^{\frac{2}{N}(k+3)} g(x-y)dxdy + 2N \int_0^{\frac{2}{N}} \int_{\frac{2}{N}(k+1)}^{\frac{2}{N}(k+2)} g(x-y)dxdy$$

$$= N \int_0^{\frac{2}{N}} \int_{\frac{2}{N}k}^{\frac{2}{N}(k+1)} \left[2g(x-y+\frac{2}{N}) - g(x-y) - g(x-y+\frac{4}{N}) \right] dxdy.$$

$$(10.32)$$

For $k \in \{2, \ldots, \frac{N}{2} - 4\}$, $y \in [0, \frac{2}{N}]$, $x \in [\frac{2}{N}k, \frac{2}{N}(k+1)]$, we have $0 < x-y < x-y+\frac{2}{N} < x-y+\frac{4}{N} < 1$, and due to the smoothness of g on $[0, 1]$ we have

$$g(x-y+\frac{2}{N}) - g(x-y) = \frac{2}{N}g'(x-y) + \frac{2}{N^2}g''(\vartheta_{k,N}(x,y))$$

$$g(x-y+\frac{4}{N}) - g(x-y+\frac{2}{N}) = \frac{2}{N}g'(x-y+\frac{2}{N}) + \frac{2}{N^2}g''(\eta_{k,N}(x,y))$$

with $\vartheta_{k,N}(x,y) \in [x-y, x-y+\frac{2}{N}] \subset [\frac{2}{N}(k-1), \frac{2}{N}(k+2)]$, $\eta_{k,N}(x,y) \in [x-y+\frac{2}{N}, x-y+\frac{4}{N}] \subset [\frac{2}{N}k, \frac{2}{N}(k+3)]$, which implies

$$|\vartheta_{k,N}(x,y) - \eta_{k,N}(x,y)| \leq \frac{8}{N}, \qquad (10.33)$$

so that by (10.32)

$$c_k^{(N)} + c_{k+2}^{(N)} - 2c_{k+1}^{(N)} = 2 \int_0^{\frac{2}{N}} \int_{\frac{2}{N}k}^{\frac{2}{N}(k+1)} \left[g'(x-y) - g'(x-y+\frac{2}{N}) \right] dxdy$$

$$+ \frac{2}{N} \int_0^{\frac{2}{N}} \int_{\frac{2}{N}k}^{\frac{2}{N}(k+1)} \left[g''(\vartheta_{k,N}(x,y)) - g''(\eta_{k,N}(x,y)) \right] dxdy$$

$$= 2 \int_0^{\frac{2}{N}} \int_{\frac{2}{N}k}^{\frac{2}{N}(k+1)} \left[g'(x-y) - g'(x-y+\frac{2}{N}) \right] dxdy + O(\frac{1}{N^4}),$$

$$(10.34)$$

since by (10.33) and the Lipschitz continuity of g'' on $[0, 1]$ (note that $\vartheta_{k,N}(x,y)$ and $\eta_{k,N}(x,y)$ lie in $[0, 1]$), we have

$$\left| \frac{2}{N} \int_0^{\frac{2}{N}} \int_{\frac{2}{N}k}^{\frac{2}{N}(k+1)} \left[g''(\vartheta_{k,N}(x,y)) - g''(\eta_{k,N}(x,y)) \right] dxdy \right|$$

$$\leq \frac{C}{N} \int_0^{\frac{2}{N}} \int_{\frac{2}{N}k}^{\frac{2}{N}(k+1)} |\vartheta_{k,N}(x,y) - \eta_{k,N}(x,y)| \leq \frac{8C}{N^2} \cdot \frac{4}{N^2} \leq \frac{C}{N^4}.$$

Thus (10.34) is valid for $k = 2, \ldots, \frac{N}{2} - 4$ even, which yields

$$\sum_{\substack{k=2 \\ k \text{ even}}}^{\frac{N}{2}-4} (c_k^{(N)} + c_{k+2}^{(N)} - 2c_{k+1}^{(N)}) \cos((k+1)\pi\delta)$$

$$= 2 \sum_{\substack{k=2 \\ k \text{ even}}}^{\frac{N}{2}-2} \int_0^{\frac{2}{N}} \int_{\frac{2}{N}k}^{\frac{2}{N}(k+1)} \left[g'(x-y) - g'(x-y+\frac{2}{N}) \right] dx dy \cos((k+1)\pi\delta) + O(\frac{1}{N^3}),$$

$$(10.35)$$

since the additional summand ($k = \frac{N}{2} - 2$) satisfies

$$\left| 2 \int_0^{\frac{2}{N}} \int_{1-\frac{4}{N}}^{1-\frac{2}{N}} \left[g'(x-y) - g'(x-y+\frac{2}{N}) \right] dx dy \cos((k+1)\pi\delta) \right| \leq \frac{C}{N^3}$$

due to the Lipschitz continuity of g', which can be applied since $[x-y, x-y+\frac{2}{N}] \subset [0,1]$. Setting (10.35) this into (10.31), we obtain

$$
\begin{aligned}
T_{t,N}^{(2)} =\ & 2\pi \frac{\delta}{N} \int_0^1 g'(y) \sin(\frac{N}{2}\pi\delta y) dy \\
& + 2 \sum_{\substack{k=2 \\ k \text{ even}}}^{\frac{N}{2}-2} \int_0^{\frac{2}{N}} \int_{\frac{2}{N}k}^{\frac{2}{N}(k+1)} \left[g'(x-y) - g'(x-y+\frac{2}{N}) \right] dx dy \cos((k+1)\pi\delta) \\
& + (c_0^{(N)} + c_2^{(N)} - 2c_1^{(N)}) \cos(\pi\delta) + (c_{\frac{N}{2}-2}^{(N)} + c_{\frac{N}{2}}^{(N)} - 2c_{\frac{N}{2}-1}^{(N)}) \cos((\frac{N}{2}-1)\pi\delta) \\
& + O(\frac{1}{N^3}) + O(\frac{\delta}{N^2}) + O(\frac{\delta^2}{N}).
\end{aligned}
$$

$$(10.36)$$

To modify the second line, we note that for $k = 2, \ldots, \frac{N}{2} - 2$, $x \in [\frac{2}{N}k, \frac{2}{N}(k+1)]$, $y \in [0, \frac{2}{N}]$ we have $[x-y, x-y+\frac{2}{N}] \subset [0,1]$, and for every such k, x, y we find $\sigma_{k,N}(x,y)$ with

$$g'(x-y) - g'(x-y+\frac{2}{N}) = -\frac{2}{N} g''(x-y) - \frac{2}{N^2} g''(\sigma_{k,N}(x,y)),$$

and $\sigma_{k,N}(x,y) \in [x-y, x-y+\frac{2}{N}] \subset [0,1]$. Thus

$$
\begin{aligned}
& 2 \sum_{\substack{k=2 \\ k \text{ even}}}^{\frac{N}{2}-2} \int_0^{\frac{2}{N}} \int_{\frac{2}{N}k}^{\frac{2}{N}(k+1)} \left[g'(x-y) - g'(x-y+\frac{2}{N}) \right] dx dy \cos((k+1)\pi\delta) = \\
& \qquad = -\frac{4}{N} \sum_{\substack{k=2 \\ k \text{ even}}}^{\frac{N}{2}-2} \int_0^{\frac{2}{N}} \int_{\frac{2}{N}k}^{\frac{2}{N}(k+1)} g''(x-y) dx dy \cos((k+1)\pi\delta) \\
& \qquad\quad - \frac{4}{N^2} \sum_{\substack{k=2 \\ k \text{ even}}}^{\frac{N}{2}-2} \int_0^{\frac{2}{N}} \int_{\frac{2}{N}k}^{\frac{2}{N}(k+1)} g''(\sigma_{k,N}(x,y)) dx dy \cos((k+1)\pi\delta) \\
& \qquad = -\frac{4}{N} \sum_{\substack{k=2 \\ k \text{ even}}}^{\frac{N}{2}-2} \int_0^{\frac{2}{N}} \int_{\frac{2}{N}k}^{\frac{2}{N}(k+1)} g''(x-y) dx dy \cos((k+1)\pi\delta) + O(\frac{1}{N^3}), \quad (10.37)
\end{aligned}
$$

since the boundedness of g'' on $[0,1]$ implies

$$\left| \frac{4}{N^2} \sum_{\substack{k=2 \\ k \text{ even}}}^{\frac{N}{2}-2} \int_0^{\frac{2}{N}} \int_{\frac{2}{N}k}^{\frac{2}{N}(k+1)} g''(\sigma_{k,N}(x,y)) dx dy \cos((k+1)\pi\delta) \right| \le \frac{C}{N^2} \sum_{\substack{k=2 \\ k \text{ even}}}^{\frac{N}{2}-2} \frac{4}{N^2} \le \frac{C}{N^3}.$$

Furthermore, for $k = 2, \ldots, \frac{N}{2} - 2$, $x \in [\frac{2}{N}k, \frac{2}{N}(k+1)]$, $y \in [0, \frac{2}{N}]$, we have $x, x - y \in [0, 1]$, where g'' is smooth so that we obtain

$$-\frac{4}{N} \sum_{\substack{k=2 \\ k \text{ even}}}^{\frac{N}{2}-2} \int_0^{\frac{2}{N}} \int_{\frac{2}{N}k}^{\frac{2}{N}(k+1)} g''(x-y) dx dy \cos((k+1)\pi\delta) =$$

$$= -\frac{4}{N} \sum_{\substack{k=2 \\ k \text{ even}}}^{\frac{N}{2}-2} \int_0^{\frac{2}{N}} \int_{\frac{2}{N}k}^{\frac{2}{N}(k+1)} g''(x) dx dy \cos((k+1)\pi\delta)$$

$$-\frac{4}{N} \sum_{\substack{k=2 \\ k \text{ even}}}^{\frac{N}{2}-2} \int_0^{\frac{2}{N}} \int_{\frac{2}{N}k}^{\frac{2}{N}(k+1)} [g''(x-y) - g''(x)] dx dy \cos((k+1)\pi\delta)$$

$$= -\frac{8}{N^2} \sum_{\substack{k=2 \\ k \text{ even}}}^{\frac{N}{2}-2} \int_{\frac{2}{N}k}^{\frac{2}{N}(k+1)} g''(x) dx \cos((k+1)\pi\delta) + O(\frac{1}{N^3}), \tag{10.38}$$

since the Lipschitz continuity of g'' on $[0,1]$ (recall that $g \in C^3[0,1]$) yields

$$\left| \frac{4}{N} \sum_{\substack{k=2 \\ k \text{ even}}}^{\frac{N}{2}-2} \int_0^{\frac{2}{N}} \int_{\frac{2}{N}k}^{\frac{2}{N}(k+1)} [g''(x-y) - g''(x)] dx dy \cos((k+1)\pi\delta) \right|$$

$$\le \frac{C}{N} \sum_{\substack{k=2 \\ k \text{ even}}}^{\frac{N}{2}-2} \int_0^{\frac{2}{N}} \int_{\frac{2}{N}k}^{\frac{2}{N}(k+1)} |y| dx dy \le \frac{2C}{N^2} \sum_{\substack{k=2 \\ k \text{ even}}}^{\frac{N}{2}-2} \int_0^{\frac{2}{N}} \int_{\frac{2}{N}k}^{\frac{2}{N}(k+1)} dx dy \le \frac{2C}{N^2} \sum_{\substack{k=2 \\ k \text{ even}}}^{\frac{N}{2}-2} \frac{4}{N^2} \le \frac{C}{N^3}.$$

Combining (10.36), (10.37) and (10.38), we obtain

$$T_{t,N}^{(2)} = 2\pi \frac{\delta}{N} \int_0^1 g'(y) \sin(\frac{N}{2}\pi\delta y) dy$$

$$-\frac{8}{N^2} \sum_{\substack{k=2 \\ k \text{ even}}}^{\frac{N}{2}-2} \int_{\frac{2}{N}k}^{\frac{2}{N}(k+1)} g''(x) dx \cos((k+1)\pi\delta)$$

$$+ (c_0^{(N)} + c_2^{(N)} - 2c_1^{(N)}) \cos(\pi\delta) + (c_{\frac{N}{2}-2}^{(N)} + c_{\frac{N}{2}}^{(N)} - 2c_{\frac{N}{2}-1}^{(N)}) \cos((\frac{N}{2} - 1)\pi\delta)$$

$$+ O(\frac{1}{N^3}) + O(\frac{\delta}{N^2}) + O(\frac{\delta^2}{N}). \tag{10.39}$$

Again, we will approximate the second line by a "Fourier-type" integral:

$$-\frac{8}{N^2} \sum_{\substack{k=2 \\ k \text{ even}}}^{\frac{N}{2}-2} \int_{\frac{2}{N}k}^{\frac{2}{N}(k+1)} g''(x)dx \cos((k+1)\pi\delta) = -\frac{8}{N^2} \sum_{\substack{k=2 \\ k \text{ even}}}^{\frac{N}{2}-2} \int_{\frac{2}{N}k}^{\frac{2}{N}(k+1)} g''(x) \cos(\frac{N}{2}\pi\delta x)dx$$

$$-\frac{8}{N^2} \sum_{\substack{k=2 \\ k \text{ even}}}^{\frac{N}{2}-2} \int_{\frac{2}{N}k}^{\frac{2}{N}(k+1)} g''(x)(\cos((k+1)\pi\delta) - \cos(\frac{N}{2}\pi\delta x))dx$$

$$= -\frac{8}{N^2} \sum_{\substack{k=2 \\ k \text{ even}}}^{\frac{N}{2}-2} \int_{\frac{2}{N}k}^{\frac{2}{N}(k+1)} g''(x) \cos(\frac{N}{2}\pi\delta x)dx + O(\frac{\delta}{N^2}), \tag{10.40}$$

since for $k = 2, \ldots, \frac{N}{2} - 2$, $x \in [\frac{2}{N}k, \frac{2}{N}(k+1)]$ we have

$$|\cos((k+1)\pi\delta) - \cos(\frac{N}{2}\pi\delta x)| \leq \pi\delta|k+1-\frac{N}{2}x| \leq \pi\delta,$$

so that, using the boundedness of g'' on $[0,1]$,

$$\left| -\frac{8}{N^2} \sum_{\substack{k=2 \\ k \text{ even}}}^{\frac{N}{2}-2} \int_{\frac{2}{N}k}^{\frac{2}{N}(k+1)} g''(x)(\cos((k+1)\pi\delta) - \cos(\frac{N}{2}\pi\delta x))dx \right|$$

$$\leq \frac{C}{N^2} \sum_{\substack{k=2 \\ k \text{ even}}}^{\frac{N}{2}-2} \int_{\frac{2}{N}k}^{\frac{2}{N}(k+1)} \pi\delta dx \leq C\frac{\delta}{N^2} \sum_{\substack{k=2 \\ k \text{ even}}}^{\frac{N}{2}-2} \frac{2}{N} \leq C\frac{\delta}{N^2}.$$

Furthermore, recalling that $\frac{N}{2}$ is even, we get

$$-\frac{8}{N^2} \sum_{\substack{k=2 \\ k \text{ even}}}^{\frac{N}{2}-2} \int_{\frac{2}{N}k}^{\frac{2}{N}(k+1)} g''(x) \cos(\frac{N}{2}\pi\delta x)dx =$$

$$= -\frac{8}{N^2} \sum_{\substack{k=0 \\ k \text{ even}}}^{\frac{N}{2}-2} \int_{\frac{2}{N}k}^{\frac{2}{N}(k+1)} g''(x) \cos(\frac{N}{2}\pi\delta x)dx + \frac{8}{N^2} \int_0^{\frac{2}{N}} g''(x) \cos(\frac{N}{2}\pi\delta x)dx$$

$$= -\frac{4}{N^2} \int_0^1 g''(x) \cos(\frac{N}{2}\pi\delta x)dx - \frac{4}{N^2} \sum_{\substack{k=0 \\ k \text{ even}}}^{\frac{N}{2}-2} \int_{\frac{2}{N}k}^{\frac{2}{N}(k+1)} g''(x) \cos(\frac{N}{2}\pi\delta x)dx$$

$$+ \frac{4}{N^2} \sum_{\substack{k=1 \\ k \text{ odd}}}^{\frac{N}{2}-1} \int_{\frac{2}{N}k}^{\frac{2}{N}(k+1)} g''(x) \cos(\frac{N}{2}\pi\delta x)dx + \frac{8}{N^2} \int_0^{\frac{2}{N}} g''(x) \cos(\frac{N}{2}\pi\delta x)dx$$

$$= -\frac{4}{N^2} \int_0^1 g''(x) \cos(\frac{N}{2}\pi\delta x)dx + \frac{8}{N^2} \int_0^{\frac{2}{N}} g''(x) \cos(\frac{N}{2}\pi\delta x)dx$$

$$+ \frac{4}{N^2} \sum_{\substack{k=0 \\ k \text{ even}}}^{\frac{N}{2}-2} \left[\int_{\frac{2}{N}(k+1)}^{\frac{2}{N}(k+2)} g''(x) \cos(\frac{N}{2}\pi\delta x)dx - \int_{\frac{2}{N}k}^{\frac{2}{N}(k+1)} g''(x) \cos(\frac{N}{2}\pi\delta x)dx \right]$$

$$= -\frac{4}{N^2} \int_0^1 g''(x) \cos(\frac{N}{2}\pi\delta x)dx + \frac{8}{N^2} \int_0^{\frac{2}{N}} g''(x) \cos(\frac{N}{2}\pi\delta x)dx$$

$$+ \frac{4}{N^2} \sum_{\substack{k=0 \\ k \text{ even}}}^{\frac{N}{2}-2} \int_{\frac{2}{N}k}^{\frac{2}{N}(k+1)} \left[g''(x + \frac{2}{N}) \cos(\frac{N}{2}\pi\delta(x + \frac{2}{N})) - g''(x) \cos(\frac{N}{2}\pi\delta x) \right] dx$$

$$= -\frac{4}{N^2} \int_0^1 g''(x) \cos(\frac{N}{2}\pi\delta x)dx + O(\frac{1}{N^3}) + O(\frac{\delta}{N^2}), \tag{10.41}$$

since

$$\left| \frac{8}{N^2} \int_0^{\frac{2}{N}} g''(x) \cos(\frac{N}{2}\pi\delta x)dx \right| \le \frac{C}{N^2} \int_0^{\frac{2}{N}} dx \le \frac{C}{N^3}$$

due to the boundedness of g'' on $[0,1]$, and since the estimate

$$\left| g''(x + \frac{2}{N}) \cos(\frac{N}{2}\pi\delta(x + \frac{2}{N})) - g''(x) \cos(\frac{N}{2}\pi\delta x) \right|$$

$$\le \left| g''(x + \frac{2}{N}) - g''(x) \right| \left| \cos(\frac{N}{2}\pi\delta(x + \frac{2}{N})) \right| + |g''(x)| \left| \cos(\frac{N}{2}\pi\delta(x + \frac{2}{N})) - \cos(\frac{N}{2}\pi\delta x) \right|$$

$$\le \frac{C}{N} + C\pi\delta,$$

which follows from the boundedness and the Lipschitz continuity of g'' on $[0,1]$, yields

$$\left| \frac{4}{N^2} \sum_{\substack{k=0 \\ k \text{ even}}}^{\frac{N}{2}-2} \int_{\frac{2}{N}k}^{\frac{2}{N}(k+1)} \left[g''(x + \frac{2}{N}) \cos(\frac{N}{2}\pi\delta(x + \frac{2}{N})) - g''(x) \cos(\frac{N}{2}\pi\delta x) \right] dx \right|$$

$$\le C(\frac{1}{N^3} + \frac{\delta}{N^2}) \sum_{\substack{k=0 \\ k \text{ even}}}^{\frac{N}{2}-2} \int_{\frac{2}{N}k}^{\frac{2}{N}(k+1)} dx \le C(\frac{1}{N^3} + \frac{\delta}{N^2}) \sum_{\substack{k=0 \\ k \text{ even}}}^{\frac{N}{2}-2} \frac{2}{N} \le C(\frac{1}{N^3} + \frac{\delta}{N^2}).$$

Combining (10.39), (10.40) and (10.41), we obtain

$$T_{t,N}^{(2)} = 2\pi \frac{\delta}{N} \int_0^1 g'(y) \sin(\frac{N}{2}\pi\delta y)dy - \frac{4}{N^2} \int_0^1 g''(x) \cos(\frac{N}{2}\pi\delta x)dx$$

$$+ (c_0^{(N)} + c_2^{(N)} - 2c_1^{(N)}) \cos(\pi\delta) + (c_{\frac{N}{2}-2}^{(N)} + c_{\frac{N}{2}}^{(N)} - 2c_{\frac{N}{2}-1}^{(N)}) \cos((\frac{N}{2} - 1)\pi\delta)$$

$$+ O(\frac{1}{N^3}) + O(\frac{\delta}{N^2}) + O(\frac{\delta^2}{N}). \tag{10.42}$$

By partial integration, we get, using $g'(1) = 0$,

$$-\frac{4}{N^2} \int_0^1 g''(x) \cos(\frac{N}{2}\pi\delta x)dx = -2\pi \frac{\delta}{N} \int_0^1 g'(x) \sin(\frac{N}{2}\pi\delta x)dx - \frac{4}{N^2} g'(x) \cos(\frac{N}{2}\pi\delta x)\Big|_0^1$$

$$= -2\pi\frac{\delta}{N}\int_0^1 g'(x)\sin(\frac{N}{2}\pi\delta x)dx + \frac{4}{N^2}g'(0).$$

Thus, (10.42) takes the form

$$T_{t,N}^{(2)} = \frac{4}{N^2}g'(0) + (c_0^{(N)} + c_2^{(N)} - 2c_1^{(N)})\cos(\pi\delta) + (c_{\frac{N}{2}-2}^{(N)} + c_{\frac{N}{2}}^{(N)} - 2c_{\frac{N}{2}-1}^{(N)})\cos((\frac{N}{2}-1)\pi\delta)$$

$$+ O(\frac{1}{N^3}) + O(\frac{\delta}{N^2}) + O(\frac{\delta^2}{N}). \tag{10.43}$$

To estimate $c_0^{(N)} + c_2^{(N)} - 2c_1^{(N)}$, one has to be careful, since the inner of the integration domain contains the point 0, in which g might not be differentiable. To get this under control, we will make use of the symmetry of g. Recalling (6.12), we observe that

$$c_0^{(N)} + c_2^{(N)} - 2c_1^{(N)} = -N\int_0^{\frac{2}{N}}\int_0^{\frac{2}{N}} g(x-y)dxdy - N\int_0^{\frac{2}{N}}\int_{\frac{4}{N}}^{\frac{6}{N}} g(x-y)dxdy$$

$$+ 2N\int_0^{\frac{2}{N}}\int_{\frac{2}{N}}^{\frac{4}{N}} g(x-y)dxdy$$

$$= N\int_0^{\frac{2}{N}}\int_0^{\frac{2}{N}}\left[2g(x-y+\frac{2}{N}) - g(x-y) - g(x-y+\frac{4}{N})\right]dxdy$$

$$= N\int_0^{\frac{2}{N}}\int_y^{\frac{2}{N}}\left[2g(x-y+\frac{2}{N}) - g(x-y) - g(x-y+\frac{4}{N})\right]dxdy$$

$$+ N\int_0^{\frac{2}{N}}\int_0^y\left[2g(x-y+\frac{2}{N}) - g(y-x) - g(x-y+\frac{4}{N})\right]dxdy. \tag{10.44}$$

Obviously, $x-y+\frac{2}{N}$, $x-y+\frac{4}{N} \in [0,1]$ for $x,y \in [0,\frac{2}{N}]$. Using the smoothness of g on $[0,1]$, for $x \leq y$ we obtain

$$-g(x-y+\frac{4}{N}) + g(x-y+\frac{2}{N}) = -\frac{2}{N}g'(x-y+\frac{2}{N}) - \frac{2}{N^2}g''(\theta_{k,N}(x,y))$$

$$g(x-y+\frac{2}{N}) - g(y-x) = \left[\frac{2}{N}+2(x-y)\right]g'(y-x) + \frac{1}{2}\left[\frac{2}{N}+2(x-y)\right]^2 g''(\gamma_{k,N}(x,y))$$

for some $\theta_{k,N}(x,y) \in [x-y+\frac{2}{N}, x-y+\frac{4}{N}] \subset [0,\frac{4}{N}]$ and some $\gamma_{k,N}(x,y) \in [0,\frac{2}{N}]$ (using $x \leq y$, one easily checks that $y-x$, $x-y+\frac{2}{N} \in [0,\frac{2}{N}]$). Thus

$$N\int_0^{\frac{2}{N}}\int_0^y\left[2g(x-y+\frac{2}{N}) - g(y-x) - g(x-y+\frac{4}{N})\right]dxdy$$

$$= 2\int_0^{\frac{2}{N}}\int_0^y\left[g'(y-x) - g'(x-y+\frac{2}{N})\right]dxdy$$

$$+ \frac{2}{N}\int_0^{\frac{2}{N}}\int_0^y\left[g''(\gamma_{k,N}(x,y)) - g''(\theta_{k,N}(x,y))\right]dxdy$$

$$+ 2N\int_0^{\frac{2}{N}}\int_0^y(x-y)g'(y-x)dxdy + 2N\int_0^{\frac{2}{N}}\int_0^y(x-y)^2 g''(\gamma_{k,N}(x,y))dxdy$$

$$+ 4 \int_0^{\frac{2}{N}} \int_0^y (x-y)g''(\gamma_{k,N}(x,y))dxdy. \tag{10.45}$$

The first line can be estimated by using the Lipschitz continuity of g' on $[0,1]$:

$$\left| 2 \int_0^{\frac{2}{N}} \int_0^y \left[g'(y-x) - g'(x-y+\frac{2}{N}) \right] dxdy \right| \le C \int_0^{\frac{2}{N}} \int_0^y \left| y-x-\frac{1}{N} \right| dxdy$$

$$\le C \int_0^{\frac{2}{N}} \int_0^y \left[|x-y| + \frac{1}{N} \right] dxdy \le \frac{C}{N} \int_0^{\frac{2}{N}} \int_0^y dxdy \le \frac{C}{N^3},$$

while the boundedness of g'' on $[0,1]$ is required for the second one:

$$\left| \frac{2}{N} \int_0^{\frac{2}{N}} \int_0^y \left[g''(\gamma_{k,N}(x,y)) - g''(\theta_{k,N}(x,y)) \right] dxdy \right| \le \frac{C}{N} \int_0^{\frac{2}{N}} \int_0^y dxdy \le \frac{C}{N^3}.$$

Furthermore, we use $|x-y| \le \frac{2}{N}$ and the boundedness of g'' to obtain

$$\left| 2N \int_0^{\frac{2}{N}} \int_0^y (x-y)^2 g''(\gamma_{k,N}(x,y))dxdy \right| \le \frac{C}{N} \int_0^{\frac{2}{N}} \int_0^y dxdy \le \frac{C}{N^3}$$

and

$$\left| 4 \int_0^{\frac{2}{N}} \int_0^y (x-y)g''(\gamma_{k,N}(x,y))dxdy \right| \le \frac{C}{N} \int_0^{\frac{2}{N}} \int_0^y dxdy \le \frac{C}{N^3}.$$

Thus, (10.45) yields

$$N \int_0^{\frac{2}{N}} \int_0^y \left[2g(x-y+\frac{2}{N}) - g(y-x) - g(x-y+\frac{4}{N}) \right] dxdy$$

$$= 2N \int_0^{\frac{2}{N}} \int_0^y (x-y)g'(y-x)dxdy + O(\frac{1}{N^3})$$

$$= 2N \int_0^{\frac{2}{N}} \int_0^y (x-y)[g'(y-x) - g'(0)]dxdy + 2N \int_0^{\frac{2}{N}} \int_0^y (x-y)g'(0)dxdy + O(\frac{1}{N^3})$$

$$= 2N \int_0^{\frac{2}{N}} \int_0^y (x-y)g'(0)dxdy + O(\frac{1}{N^3}), \tag{10.46}$$

since by the Lipschitz continuity of g' on $[0,1]$ and the estimate $|x-y| \le \frac{2}{N}$,

$$\left| 2N \int_0^{\frac{2}{N}} \int_0^y (x-y)[g'(y-x) - g'(0)]dxdy \right| \le CN \int_0^{\frac{2}{N}} \int_0^y |x-y|^2 dxdy$$

$$\le \frac{C}{N} \int_0^{\frac{2}{N}} \int_0^y dxdy \le \frac{C}{N^3}.$$

A simple calculation yields

$$2N \int_0^{\frac{2}{N}} \int_0^y (x-y)g'(0)dxdy = -\frac{8}{3N^2}g'(0),$$

so that by (10.46)

$$N \int_0^{\frac{2}{N}} \int_0^y \left[2g(x-y+\frac{2}{N}) - g(y-x) - g(x-y+\frac{4}{N}) \right] dxdy = -\frac{8}{3N^2}g'(0) + O(\frac{1}{N^3}). \quad (10.47)$$

As for the first line in (10.44), we can go on as usual: Since $[x-y, x-y+\frac{4}{N}] \subset [0,1]$ due to $x \geq y$, we can make use of the smoothness of g to obtain

$$g(x-y+\frac{2}{N}) - g(x-y) = \frac{2}{N}g'(x-y) + \frac{2}{N^2}g''(\tilde{\theta}_{k,N}(x,y))$$

$$g(x-y+\frac{2}{N}) - g(x-y+\frac{4}{N}) = -\frac{2}{N}g'(x-y+\frac{2}{N}) - \frac{2}{N^2}g''(\tilde{\gamma}_{k,N}(x,y))$$

with $\tilde{\theta}_{k,N}(x,y), \tilde{\gamma}_{k,N}(x,y) \in [0,1]$, so that

$$N \int_0^{\frac{2}{N}} \int_y^{\frac{2}{N}} \left[2g(x-y+\frac{2}{N}) - g(x-y) - g(x-y+\frac{4}{N}) \right] dxdy =$$

$$= 2 \int_0^{\frac{2}{N}} \int_y^{\frac{2}{N}} \left[g'(x-y) - g'(x-y+\frac{2}{N}) \right] dxdy$$

$$+ \frac{2}{N} \int_0^{\frac{2}{N}} \int_y^{\frac{2}{N}} \left[g''(\tilde{\theta}_{k,N}(x,y)) - g''(\tilde{\gamma}_{k,N}(x,y)) \right] dxdy = O(\frac{1}{N^3}), \quad (10.48)$$

which follows from the Lipschitz continuity of g':

$$\left| 2 \int_0^{\frac{2}{N}} \int_y^{\frac{2}{N}} \left[g'(x-y) - g'(x-y+\frac{2}{N}) \right] dxdy \right| \leq \frac{C}{N} \int_0^{\frac{2}{N}} \int_y^{\frac{2}{N}} dxdy \leq \frac{C}{N^3},$$

while the boundedness of g'' on $[0,1]$ implies

$$\left| \frac{2}{N} \int_0^{\frac{2}{N}} \int_y^{\frac{2}{N}} \left[g''(\tilde{\theta}_{k,N}(x,y)) - g''(\tilde{\gamma}_{k,N}(x,y)) \right] dxdy \right| \leq \frac{C}{N} \int_0^{\frac{2}{N}} \int_y^{\frac{2}{N}} dxdy \leq \frac{C}{N^3}.$$

Combining (10.47) and (10.48) with (10.44), we obtain

$$c_0^{(N)} + c_2^{(N)} - 2c_1^{(N)} = -\frac{8}{3N^2}g'(0) + O(\frac{1}{N^3}),$$

so that (10.43) becomes

$$T_{l,N}^{(2)} = \frac{4}{N^2}g'(0) - \frac{8}{3N^2}g'(0)\cos(\pi\delta) + (c_{\frac{N}{2}-2}^{(N)} + c_{\frac{N}{2}}^{(N)} - 2c_{\frac{N}{2}-1}^{(N)})\cos((\frac{N}{2}-1)\pi\delta)$$

$$+ O(\frac{1}{N^3}) + O(\frac{\delta}{N^2}) + O(\frac{\delta^2}{N})$$

$$= \frac{4}{N^2}g'(0) - \frac{8}{3N^2}g'(0) + \frac{8}{3N^2}g'(0)[\cos 0 - \cos(\pi\delta)]$$

$$+ (c_{\frac{N}{2}-2}^{(N)} + c_{\frac{N}{2}}^{(N)} - 2c_{\frac{N}{2}-1}^{(N)})\cos((\frac{N}{2}-1)\pi\delta)$$

$$+ O(\frac{1}{N^3}) + O(\frac{\delta}{N^2}) + O(\frac{\delta^2}{N})$$

$$= \frac{4}{3N^2} g'(0) + (c^{(N)}_{\frac{N}{2}-2} + c^{(N)}_{\frac{N}{2}} - 2c^{(N)}_{\frac{N}{2}-1}) \cos((\frac{N}{2} - 1)\pi\delta)$$

$$+ O(\frac{1}{N^3}) + O(\frac{\delta}{N^2}) + O(\frac{\delta^2}{N}) \tag{10.49}$$

since

$$\left| \frac{8}{3N^2} g'(0)[\cos 0 - \cos(\pi\delta)] \right| \le \frac{C}{N^2}\pi\delta = O(\frac{\delta}{N^2}).$$

We still have to compute $c^{(N)}_{\frac{N}{2}-2} + c^{(N)}_{\frac{N}{2}} - 2c^{(N)}_{\frac{N}{2}-1}$. To do so, we recall (6.12) and the fact that $g(2 - x) = g(x - 2) = g(x)$ for every $x \in \mathbb{R}$ to obtain

$$c^{(N)}_{\frac{N}{2}-2} + c^{(N)}_{\frac{N}{2}} - 2c^{(N)}_{\frac{N}{2}-1} = -N \int_0^{\frac{2}{N}} \int_{1-\frac{4}{N}}^{1-\frac{2}{N}} g(x - y)dxdy - N \int_0^{\frac{2}{N}} \int_1^{1+\frac{2}{N}} g(x - y)dxdy$$

$$+ 2N \int_0^{\frac{2}{N}} \int_{1-\frac{2}{N}}^1 g(x - y)dxdy$$

$$= N \int_0^{\frac{2}{N}} \int_{1-\frac{2}{N}}^1 \left[2g(x - y) - g(x - y + \frac{2}{N}) - g(x - y - \frac{2}{N}) \right]dxdy$$

$$= N \int_0^{\frac{2}{N}} \int_{1-\frac{2}{N}}^{1-\frac{2}{N}+y} \left[2g(x - y) - g(x - y + \frac{2}{N}) - g(x - y - \frac{2}{N}) \right]dxdy$$

$$+ N \int_0^{\frac{2}{N}} \int_{1-\frac{2}{N}+y}^1 \left[2g(x - y) - g(2 - \frac{2}{N} - x + y) - g(x - y - \frac{2}{N}) \right]dxdy. \tag{10.50}$$

To modify the second integral, we make sure that $x - y,\ x - y - \frac{2}{N},\ 2 - \frac{2}{N} - x + y \in [0, 1]$ for $y \in [0, \frac{2}{N}]$, $x \in [1 - \frac{2}{N} + y, 1]$, and we obtain

$$g(x - y) - g(x - y - \frac{2}{N}) = -\frac{2}{N}g'(x - y) - \frac{2}{N^2}g''(\xi_{k,N}(x, y))$$

$$g(x - y) - g(2 - \frac{2}{N} - x + y) = 2(x - y - 1 + \frac{1}{N})g'(2 - \frac{2}{N} - x + y)$$

$$+ 2(x - y - 1 + \frac{1}{N})^2 g''(\eta_{k,N}(x, y))$$

for some $\xi_{k,N}(x, y),\ \eta_{k,N}(x, y)$ in $[0, 1]$, so that

$$N \int_0^{\frac{2}{N}} \int_{1-\frac{2}{N}+y}^1 \left[2g(x - y) - g(2 - \frac{2}{N} - x + y) - g(x - y - \frac{2}{N}) \right]dxdy$$

$$= 2 \int_0^{\frac{2}{N}} \int_{1-\frac{2}{N}+y}^1 \left[g'(2 - \frac{2}{N} - x + y) - g'(x - y) \right]dxdy$$

$$+ \frac{2}{N} \int_0^{\frac{2}{N}} \int_{1-\frac{2}{N}+y}^1 \left[g''(\eta_{k,N}(x, y)) - g''(\xi_{k,N}(x, y)) \right]dxdy$$

$$+ 2N \int_0^{\frac{2}{N}} \int_{1-\frac{2}{N}+y}^1 (x-y-1)g'(2-\frac{2}{N}-x+y)dxdy$$

$$+ 2N \int_0^{\frac{2}{N}} \int_{1-\frac{2}{N}+y}^1 (x-y-1)^2 g''(\eta_{k,N}(x,y))dxdy$$

$$+ 4 \int_0^{\frac{2}{N}} \int_{1-\frac{2}{N}+y}^1 (x-y-1)g''(\eta_{k,N}(x,y))dxdy. \tag{10.51}$$

To estimate the first line, we use the Lipschitz continuity of g' on $[0,1]$:

$$\left| 2 \int_0^{\frac{2}{N}} \int_{1-\frac{2}{N}+y}^1 \left[g'(2-\frac{2}{N}-x+y) - g'(x-y) \right] dxdy \right|$$

$$\leq C \int_0^{\frac{2}{N}} \int_{1-\frac{2}{N}+y}^1 \left| 1 - \frac{1}{N} - x + y \right| dxdy \leq C \int_0^{\frac{2}{N}} \int_{1-\frac{2}{N}+y}^1 \left[\frac{1}{N} + |1-x+y| \right] dxdy$$

$$= C \int_0^{\frac{2}{N}} \int_{1-\frac{2}{N}+y}^1 \left[\frac{1}{N} + 1 - x + y \right] dxdy \leq \frac{3C}{N} \int_0^{\frac{2}{N}} \int_{1-\frac{2}{N}+y}^1 dxdy \leq \frac{C}{N^3}.$$

For the second line, we easily deduce from the boundedness of g''

$$\left| \frac{2}{N} \int_0^{\frac{2}{N}} \int_{1-\frac{2}{N}+y}^1 \left[g''(\eta_{k,N}(x,y)) - g''(\xi_{k,N}(x,y)) \right] dxdy \right| \leq \frac{C}{N} \int_0^{\frac{2}{N}} \int_{1-\frac{2}{N}+y}^1 dxdy \leq \frac{C}{N^3}.$$

To consider the fourth and fifth line, we use the fact that for $y \in [0, \frac{2}{N}]$, $x \in [1-\frac{2}{N}+y, 1]$, we have $|x-y-1| = 1 - x + y \leq \frac{2}{N}$ and the boundedness of g'' to obtain

$$\left| 2N \int_0^{\frac{2}{N}} \int_{1-\frac{2}{N}+y}^1 (x-y-1)^2 g''(\eta_{k,N}(x,y))dxdy \right| \leq 2N \int_0^{\frac{2}{N}} \int_{1-\frac{2}{N}+y}^1 \frac{C}{N^2} dxdy \leq \frac{C}{N^3}$$

and

$$\left| 4 \int_0^{\frac{2}{N}} \int_{1-\frac{2}{N}+y}^1 (x-y-1)g''(\eta_{k,N}(x,y))dxdy \right| \leq 4 \int_0^{\frac{2}{N}} \int_{1-\frac{2}{N}+y}^1 \frac{C}{N} dxdy \leq \frac{C}{N^3}.$$

For the third line, we derive, using $g'(1) = 0$,

$$2N \int_0^{\frac{2}{N}} \int_{1-\frac{2}{N}+y}^1 (x-y-1)g'(2-\frac{2}{N}-x+y)dxdy$$

$$= 2N \int_0^{\frac{2}{N}} \int_{1-\frac{2}{N}+y}^1 (x-y-1) \left[g'(2-\frac{2}{N}-x+y) - g'(1) \right] dxdy = O(\frac{1}{N^3})$$

since due to the Lipschitz continuity of g' on $[0,1]$ and the estimate $|x-y-1| \leq \frac{2}{N}$

$$\left| 2N \int_0^{\frac{2}{N}} \int_{1-\frac{2}{N}+y}^1 (x-y-1) \left[g'(2-\frac{2}{N}-x+y) - g'(1) \right] dxdy \right|$$

$$\leq CN \int_0^{\frac{2}{N}} \int_{1-\frac{2}{N}+y}^1 \left| x-y-1 \right| \left| 1 - \frac{2}{N} - x + y \right| dx dy \leq \frac{C}{N} \int_0^{\frac{2}{N}} \int_{1-\frac{2}{N}+y}^1 dx dy \leq \frac{C}{N^3}.$$

(10.51) now implies

$$N \int_0^{\frac{2}{N}} \int_{1-\frac{2}{N}+y}^1 \left[2g(x-y) - g(2 - \frac{2}{N} - x + y) - g(x - y - \frac{2}{N}) \right] dx dy = O(\frac{1}{N^3}), \quad (10.52)$$

while the estimate

$$N \int_0^{\frac{2}{N}} \int_{1-\frac{2}{N}}^{1-\frac{2}{N}+y} \left[2g(x-y) - g(x - y + \frac{2}{N}) - g(x - y - \frac{2}{N}) \right] dx dy = O(\frac{1}{N^3}) \quad (10.53)$$

is shown analogously to (10.48), and setting (10.53) and (10.52) into (10.50) we obtain

$$c_{\frac{N}{2}-2}^{(N)} + c_{\frac{N}{2}}^{(N)} - 2c_{\frac{N}{2}-1}^{(N)} = O(\frac{1}{N^3}),$$

so that (10.49) becomes

$$T_{t,N}^{(2)} = \frac{4}{3N^2} g'(0) + O(\frac{1}{N^3}) + O(\frac{\delta}{N^2}) + O(\frac{\delta^2}{N}),$$

which completes the proof of Lemma 10.3. $\qquad\qquad\qquad\qquad\qquad\qquad\qquad$ $\square$

From (10.6), Lemma 10.1 and 10.3 we immediately deduce

Lemma 10.4 *Let $N \in \mathbb{N}$ be a multiple of 4 (i.e., $\frac{N}{2}$ is even), $t \in \{1, \ldots, \frac{N}{2} - 1\}$, δ as given by (10.9). Then*

$$c_0^{(N)} + (-1)^t c_{\frac{N}{2}}^{(N)} + 2 \sum_{k=1}^{\frac{N}{2}-1} c_k^{(N)} \cos\left(\frac{2\pi kt}{N}\right) = \frac{4}{3N^2} g'(0) + O(\delta^3) + O(\frac{\delta^2}{N}) + O(\frac{\delta}{N^2}) + O(\frac{1}{N^3}).$$

10.3   Positivity of Eigenvalues

Now that we have computed the "middle part" in (10.5), we will pass over to the estimation of the other sum:

Lemma 10.5 *Let $N \in \mathbb{N}$ be a multiple of 4 (i.e., $\frac{N}{2}$ is even), $t \in \{1, \ldots, \frac{N}{2} - 1\}$, δ as given by (10.9). Then*

$$2N \sum_{k=0}^{\frac{N}{2}-1} (-1)^k \int_0^{\frac{1}{N}} x \left[g(\frac{2k+1}{N} - x) - g(\frac{2k+1}{N} + x) \right] dx \left(1 - \cos(\frac{2\pi t}{N}) \right)$$

$$= -\frac{4}{3N^2} g'(0) + O(\frac{\delta}{N^2}) + O(\frac{1}{N^3})$$

Proof: Since $\frac{N}{2}$ is even, we can rewrite the alternating sum as

$$2N \sum_{k=0}^{\frac{N}{2}-1} (-1)^k \int_0^{\frac{1}{N}} x \Big[g(\frac{2k+1}{N} - x) - g(\frac{2k+1}{N} + x) \Big] dx$$

$$= 2N \sum_{\substack{k=0 \\ k \text{ even}}}^{\frac{N}{2}-2} \int_0^{\frac{1}{N}} x \Big[g(\frac{2k+1}{N} - x) - g(\frac{2k+1}{N} + x) - g(\frac{2k+3}{N} - x) + g(\frac{2k+3}{N} + x) \Big] dx$$

$$= 2N \sum_{\substack{k=0 \\ k \text{ even}}}^{\frac{N}{2}-2} \int_0^{\frac{1}{N}} x \Big[F_x(\frac{2k+1}{N}) - F_x(\frac{2k+3}{N}) \Big] dx,$$

where $F_x(y) = g(y - x) - g(y + x)$. After checking that $[y - x, y + x] \subset [0, 1]$ for every $x \in [0, \frac{1}{N}]$, $y \in [\frac{2k+1}{N}, \frac{2k+3}{N}]$ for $k = 0, \dots, \frac{N}{2} - 2$, we can make use of the smoothness of g to find $\zeta_{k,N}(x) \in [\frac{2k+1}{N}, \frac{2k+3}{N}]$ such that

$$F_x(\frac{2k+1}{N}) - F_x(\frac{2k+3}{N}) = -\frac{2}{N} F_x'(\frac{2k+1}{N}) - \frac{2}{N^2} F_x''(\zeta_{k,N}(x))$$

$$= -\frac{2}{N} g'(\frac{2k+1}{N} - x) + \frac{2}{N} g'(\frac{2k+1}{N} + x)$$

$$- \frac{2}{N^2} g''(\zeta_{k,N}(x) - x) + \frac{2}{N^2} g''(\zeta_{k,N}(x) + x).$$

Thus

$$2N \sum_{k=0}^{\frac{N}{2}-1} (-1)^k \int_0^{\frac{1}{N}} x \Big[g(\frac{2k+1}{N} - x) - g(\frac{2k+1}{N} + x) \Big] dx$$

$$= 4 \sum_{\substack{k=0 \\ k \text{ even}}}^{\frac{N}{2}-2} \int_0^{\frac{1}{N}} x \Big[g'(\frac{2k+1}{N} + x) - g'(\frac{2k+1}{N} - x) \Big] dx$$

$$+ \frac{4}{N} \sum_{\substack{k=0 \\ k \text{ even}}}^{\frac{N}{2}-2} \int_0^{\frac{1}{N}} x \Big[g''(\zeta_{k,N}(x) + x) - g''(\zeta_{k,N}(x) - x) \Big] dx$$

$$= 4 \sum_{\substack{k=0 \\ k \text{ even}}}^{\frac{N}{2}-2} \int_0^{\frac{1}{N}} x \Big[g'(\frac{2k+1}{N} + x) - g'(\frac{2k+1}{N} - x) \Big] dx + O(\frac{1}{N^3}), \quad (10.54)$$

since the Lipschitz continuity of g'' on $[0, 1]$ implies

$$\left| \frac{4}{N} \sum_{\substack{k=0 \\ k \text{ even}}}^{\frac{N}{2}-2} \int_0^{\frac{1}{N}} x \Big[g''(\zeta_{k,N}(x) + x) - g''(\zeta_{k,N}(x) - x) \Big] dx \right|$$

$$\leq \frac{C}{N} \sum_{\substack{k=0 \\ k \text{ even}}}^{\frac{N}{2}-2} \int_0^{\frac{1}{N}} 2x^2 dx \leq \frac{C}{N^3} \sum_{\substack{k=0 \\ k \text{ even}}}^{\frac{N}{2}-2} \int_0^{\frac{1}{N}} dx \leq \frac{C}{N^3}.$$

We use the property $g \in C^3[0,1]$ to find $\xi_{k,N}(x) \in [0,1]$ ($k = 0, \ldots, \frac{N}{2} - 2$ even) with

$$g'(\frac{2k+1}{N} + x) - g'(\frac{2k+1}{N} - x) = 2xg''(\frac{2k+1}{N} - x) + 2x^2g'''(\xi_{k,N}(x))$$

which set into (10.54) yields

$$2N \sum_{k=0}^{\frac{N}{2}-1} (-1)^k \int_0^{\frac{1}{N}} x\Big[g(\frac{2k+1}{N} - x) - g(\frac{2k+1}{N} + x)\Big] dx$$

$$= 8 \sum_{\substack{k=0 \\ k \text{ even}}}^{\frac{N}{2}-2} \int_0^{\frac{1}{N}} x^2 g''(\frac{2k+1}{N} - x) dx + 8 \sum_{\substack{k=0 \\ k \text{ even}}}^{\frac{N}{2}-2} \int_0^{\frac{1}{N}} x^3 g'''(\xi_{k,N}(x)) dx + O(\frac{1}{N^3})$$

$$= 8 \sum_{\substack{k=0 \\ k \text{ even}}}^{\frac{N}{2}-2} \int_0^{\frac{1}{N}} x^2 g''(\frac{2k+1}{N} - x) dx + O(\frac{1}{N^3}), \tag{10.55}$$

since the boundedness of g''' implies

$$\Big| 8 \sum_{\substack{k=0 \\ k \text{ even}}}^{\frac{N}{2}-2} \int_0^{\frac{1}{N}} x^3 g'''(\xi_{k,N}(x)) dx \Big| \leq C \sum_{\substack{k=0 \\ k \text{ even}}}^{\frac{N}{2}-2} \int_0^{\frac{1}{N}} \frac{1}{N^3} dx \leq \frac{C}{N^3}.$$

Furthermore,

$$8 \sum_{\substack{k=0 \\ k \text{ even}}}^{\frac{N}{2}-2} \int_0^{\frac{1}{N}} x^2 g''(\frac{2k+1}{N} - x) dx =$$

$$= 8 \sum_{\substack{k=0 \\ k \text{ even}}}^{\frac{N}{2}-2} \int_0^{\frac{1}{N}} x^2 g''(\frac{2k}{N}) dx + 8 \sum_{\substack{k=0 \\ k \text{ even}}}^{\frac{N}{2}-2} \int_0^{\frac{1}{N}} x^2\Big[g''(\frac{2k+1}{N} - x) - g''(\frac{2k}{N})\Big] dx$$

$$= 8 \sum_{\substack{k=0 \\ k \text{ even}}}^{\frac{N}{2}-2} \int_0^{\frac{1}{N}} x^2 g''(\frac{2k}{N}) dx + O(\frac{1}{N^3}),$$

since from the Lipschitz continuity of g'' on $[0,1]$ we deduce

$$\Big| 8 \sum_{\substack{k=0 \\ k \text{ even}}}^{\frac{N}{2}-2} \int_0^{\frac{1}{N}} x^2\Big[g''(\frac{2k+1}{N} - x) - g''(\frac{2k}{N})\Big] dx \Big| \leq C \sum_{\substack{k=0 \\ k \text{ even}}}^{\frac{N}{2}-2} \int_0^{\frac{1}{N}} \frac{1}{N^2}\Big|\frac{1}{N} - x\Big| dx \leq \frac{C}{N^3}.$$

Thus, from (10.55) we obtain, recalling that $\frac{N}{2}$ is even,

$$2N \sum_{k=0}^{\frac{N}{2}-1} (-1)^k \int_0^{\frac{1}{N}} x\Big[g(\frac{2k+1}{N} - x) - g(\frac{2k+1}{N} + x)\Big] dx$$

$$=8\sum_{\substack{k=0\\k\ even}}^{\frac{N}{2}-2}\int_0^{\frac{1}{N}}x^2g''(\frac{2k}{N})dx+O(\frac{1}{N^3})=\frac{8}{3N^3}\sum_{\substack{k=0\\k\ even}}^{\frac{N}{2}-2}g''(\frac{2k}{N})+O(\frac{1}{N^3})$$

$$=\frac{4}{3N^3}\sum_{k=0}^{\frac{N}{2}-1}g''(\frac{2k}{N})+\frac{4}{3N^3}\sum_{\substack{k=0\\k\ even}}^{\frac{N}{2}-1}g''(\frac{2k}{N})-\frac{4}{3N^3}\sum_{\substack{k=1\\k\ odd}}^{\frac{N}{2}-1}g''(\frac{2k}{N})+O(\frac{1}{N^3})$$

$$=\frac{4}{3N^3}\sum_{k=0}^{\frac{N}{2}-1}g''(\frac{2k}{N})+\frac{4}{3N^3}\sum_{\substack{k=0\\k\ even}}^{\frac{N}{2}-2}\left[g''(\frac{2k}{N})-g''(\frac{2k+2}{N})\right]+O(\frac{1}{N^3})$$

$$=\frac{2}{3N^2}\sum_{k=0}^{\frac{N}{2}-1}\frac{2}{N}g''(\frac{2k}{N})+O(\frac{1}{N^3}),\tag{10.56}$$

which follows from the Lipschitz continuity of g'' on $[0,1]$ as follows:

$$\left|\frac{4}{3N^3}\sum_{\substack{k=0\\k\ even}}^{\frac{N}{2}-2}\left[g''(\frac{2k}{N})-g''(\frac{2k+2}{N})\right]\right|\le\frac{C}{N^3}\sum_{\substack{k=0\\k\ even}}^{\frac{N}{2}-2}\frac{2}{N}\le\frac{C}{N^3}.$$

We approximate the remaining sum by an integral:

$$\frac{2}{3N^2}\sum_{k=0}^{\frac{N}{2}-1}\frac{2}{N}g''(\frac{2k}{N})=\frac{2}{3N^2}\sum_{k=0}^{\frac{N}{2}-1}\int_{\frac{2}{N}k}^{\frac{2}{N}(k+1)}g''(x)dx+\frac{2}{3N^2}\sum_{k=0}^{\frac{N}{2}-1}\int_{\frac{2}{N}k}^{\frac{2}{N}(k+1)}\left[g''(\frac{2k}{N})-g''(x)\right]dx$$

$$=\frac{2}{3N^2}\sum_{k=0}^{\frac{N}{2}-1}\int_{\frac{2}{N}k}^{\frac{2}{N}(k+1)}g''(x)dx+O(\frac{1}{N^3}),$$

since

$$\left|\frac{2}{3N^2}\sum_{k=0}^{\frac{N}{2}-1}\int_{\frac{2}{N}k}^{\frac{2}{N}(k+1)}\left[g''(\frac{2k}{N})-g''(x)\right]dx\right|\le\frac{C}{N^2}\sum_{k=0}^{\frac{N}{2}-1}\int_{\frac{2}{N}k}^{\frac{2}{N}(k+1)}\left|x-\frac{2k}{N}\right|dx$$

$$\le\frac{C}{N^2}\sum_{k=0}^{\frac{N}{2}-1}\int_{\frac{2}{N}k}^{\frac{2}{N}(k+1)}\frac{2}{N}dx\le\frac{C}{N^2}\sum_{k=0}^{\frac{N}{2}-1}\frac{4}{N^2}\le\frac{C}{N^3}$$

(10.56) now becomes

$$2N\sum_{k=0}^{\frac{N}{2}-1}(-1)^k\int_0^{\frac{1}{N}}x\left[g(\frac{2k+1}{N}-x)-g(\frac{2k+1}{N}+x)\right]dx$$

$$=\frac{2}{3N^2}\sum_{k=0}^{\frac{N}{2}-1}\int_{\frac{2}{N}k}^{\frac{2}{N}(k+1)}g''(x)dx+O(\frac{1}{N^3})=\frac{2}{3N^2}\int_0^1g''(x)dx+O(\frac{1}{N^3})$$

$$=-\frac{2}{3N^2}g'(0)+O(\frac{1}{N^3})\tag{10.57}$$

since $g'(1) = 0$. Recalling (10.9), we deduce

$$1 - \cos(\frac{2\pi t}{N}) = 1 - \cos(\pi(1 - \delta)) = 1 + \cos(\pi\delta) = 2 + [\cos(\pi\delta) - \cos(0)] = 2 + O(\delta),$$

which together with (10.57) yields

$$2N \sum_{k=0}^{\frac{N}{2}-1} (-1)^k \int_0^{\frac{1}{N}} x \left[g(\frac{2k+1}{N} - x) - g(\frac{2k+1}{N} + x) \right] dx \left(1 - \cos(\frac{2\pi t}{N})\right)$$

$$= \left[-\frac{2}{3N^2} g'(0) + O(\frac{1}{N^3}) \right] (2 + O(\delta)) = -\frac{4}{3N^2} g'(0) + O(\frac{1}{N^3}) + O(\frac{\delta}{N^2}),$$

which is what we wanted to show.   □

We can put these pieces together now to get an estimate for the eigenvalues:

Theorem 10.6 *Let $N \in \mathbb{N}$ be a multiple of 4 (i.e., $\frac{N}{2}$ is even), $t \in \{1, \ldots, \frac{N}{2} - 1\}$, δ as given by (10.9). Then*

$$\lambda_t^{(N)} = \frac{\pi^2}{2} \delta^2 \int_{-1}^1 g(y) dy + O(\frac{1}{N^3}) + O(\frac{\delta}{N^2}) + O(\frac{\delta^2}{N}) + O(\delta^3)$$

Proof: From (10.5) we deduce, using Lemma 10.4 and Lemma 10.5,

$$\lambda_t^{(N)} = \int_{-1}^1 g(y) dy \left(1 + \cos(\frac{2\pi t}{N})\right) + \frac{4}{3N^2} g'(0) + O(\delta^3) + O(\frac{\delta^2}{N}) + O(\frac{\delta}{N^2}) + O(\frac{1}{N^3})$$

$$- \frac{4}{3N^2} g'(0) + O(\frac{\delta}{N^2}) + O(\frac{1}{N^3})$$

$$= \int_{-1}^1 g(y) dy \left(1 + \cos(\frac{2\pi t}{N})\right) + O(\delta^3) + O(\frac{\delta^2}{N}) + O(\frac{\delta}{N^2}) + O(\frac{1}{N^3}). \qquad (10.58)$$

Using (10.9), we obtain

$$1 + \cos(\frac{2\pi t}{N}) = \cos(0) + \cos(\pi(1 - \delta)) = \cos(0) - \cos(\pi\delta) = \frac{\pi^2}{2} \delta^2 + O(\delta^4)$$

by expanding cos around 0, which together with (10.58) implies the assertion.   □

Now, the positivity of eigenvalues for large multiples of 4 is an almost immediate consequence of the previous theorem:

Corollary 10.7 *There exist $\mu_0 > 0$, $N_0 \in \mathbb{N}$ and $\delta_0 > 0$ such that for every $N \geq N_0$ which is a multiple of 4, the following condition is satisfied:*

For every $t \in \{1, \ldots, \frac{N}{2} - 1\}$ satisfying

$$\frac{t}{N} \geq \frac{1}{2} - \delta_0,$$

we have the estimate

$$\lambda_t^{(N)} \geq \frac{\mu_0}{N^2}.$$

Proof: For $t \in \{1, \ldots, \frac{N}{2} - 1\}$, we have

$$\delta = \delta_{t,N} = 1 - \frac{2t}{N} \geq 1 - \frac{2}{N}(\frac{N}{2} - 1) = \frac{2}{N}, \tag{10.59}$$

so that by Theorem 10.6 we find a constant $C > 0$ such that

$$\lambda_t^{(N)} \geq \frac{\pi^2}{N^2} \int_{-1}^{1} g(y)dy + O(\frac{1}{N^3}) + O(\frac{\delta}{N^2}) + \frac{\pi^2}{4}\delta^2 \int_{-1}^{1} g(y)dy + O(\frac{\delta^2}{N}) + O(\delta^3)$$

$$\geq \frac{\pi^2}{N^2} \int_{-1}^{1} g(y)dy \left[1 - \frac{C}{N} - C\delta\right] + \frac{\pi^2}{4}\delta^2 \int_{-1}^{1} g(y)dy \left[1 - \frac{C}{N} - C\delta\right]$$

Thus we find $N_0 \in \mathbb{N}$ and $\delta_0 > 0$ such that for every $N \geq N_0$ multiple of 4 and for $\delta \leq 2\delta_0$ we get

$$\lambda_t^{(N)} \geq \frac{\pi^2}{2N^2} \int_{-1}^{1} g(y)dy + \frac{\pi^2}{8}\delta^2 \int_{-1}^{1} g(y)dy \geq \frac{\pi^2}{N^2} \int_{-1}^{1} g(y)dy =: \frac{\mu_0}{N^2}.$$

where we made use of (10.59). The condition $\delta \leq 2\delta_0$, however, by (10.9) is equivalent to

$$1 - \frac{2t}{N} \leq 2\delta_0, \quad \text{i.e.} \quad \frac{t}{N} \geq \frac{1}{2} - \delta_0,$$

which proves Theorem 10.7. $\qquad\qquad\qquad\qquad\qquad\qquad\qquad\qquad\qquad\qquad\qquad\qquad\square$

Corollaries 9.6 and 10.7 imply the following positivity condition:

Corollary 10.8 *There exist $\mu_0 > 0$ and $N_0 \in \mathbb{N}$ such that for every $N \geq N_0$ which is a multiple of 4, we have*

$$\lambda_t^{(N)} \geq \frac{\mu_0}{N^2}$$

for every $t \in \{1, \ldots, \frac{N}{2} - 1\}$. In particular, for every $N \geq N_0$ which is a multiple of 4, $\lambda_t^{(N)} > 0$ for $t = 1, \ldots, \frac{N}{2} - 1$.

Proof: We apply Corollary 10.7 to obtain $\mu_0 > 0$, $N_1 \in \mathbb{N}$ and $\delta_0 > 0$ such that for every $N \geq N_1$ which is a multiple of 4, we have

$$\lambda_t^{(N)} \geq \frac{\mu_0}{N^2} \quad \text{for every} \quad t \in \{1, \ldots, \frac{N}{2} - 1\} \quad \text{with} \quad \frac{t}{N} \geq \frac{1}{2} - \delta_0. \tag{10.60}$$

Using Corollary 9.6, for δ_0 we find $N_2 \in \mathbb{N}$ and $\nu_0 > 0$ such that for every $N \geq N_2$, the estimate

$$\lambda_t^{(N)} \geq \nu_0 \quad \text{for every} \quad t \in \{1, \ldots, \frac{N}{2} - 1\} \quad \text{with} \quad \frac{t}{N} \leq \frac{1}{2} - \delta_0 \tag{10.61}$$

holds true. We choose $N_3 \in \mathbb{N}$ such that

$$\nu_0 \geq \frac{\mu_0}{N^2} \quad \text{for every} \quad N \geq N_3,$$

then for every $N \geq N_0 := \max\{N_1, N_2, N_3\}$ which is a multiple of 4, we obtain, combining (10.60) and (10.61),

$$\lambda_t^{(N)} \geq \frac{\mu_0}{N^2} \quad \text{for every} \quad t \in \{1, \ldots, \frac{N}{2} - 1\}.$$

$\square$

10.4 Proof of the Main Theorem

The main theorem now follows immediately:

Proof of Theorem 2.3: Due to Corollary 10.8, we find $N_0 \in \mathbb{N}$ such that for every $N \geq N_0$ which is a multiple of 4, we have

$$\lambda_t^{(N)} > 0 \quad \text{for} \quad t = 1, \ldots \frac{N}{2} - 1.$$

Application of Corollary 8.6 (iii) yields that for every $N \in \mathbb{N}$, $N \geq N_0$ which is a multiple of 4, we find $\varepsilon_0 > 0$ and $\delta > 0$ such that for every $\varepsilon \in (0, \varepsilon_0)$, we find an H^1-local minimizer $u_\varepsilon \in H_{per}^{2,0}(-1,1)$ of I^ε with

$$I^\varepsilon(u_\varepsilon) \leq I^\varepsilon(u) \quad \text{for every} \quad u \in H_{per}^{2,0}(-1,1) \quad \text{with} \quad \|u - \bar{u}_N\|_{H^1(-1,1)} < \delta$$

and

$$\lim_{\varepsilon \to 0} \|u_\varepsilon - \bar{u}_N\|_{H^1(-1,1)} = 0,$$

which shows Theorem 2.3.

$\square$

10.5 The Case $\frac{N}{2}$ Odd

I have mentioned that Theorem 2.3 also holds if we drop the condition that $\frac{N}{2}$ is even. Obviously, this would immediately follow if we are able to prove that all the estimates above also hold if $\frac{N}{2}$ is odd. We only sketch the proof in order to show the modifications required, while other steps are done analogously to the estimates done in the even case. First of all, we recall (10.7), which can be written as

$$T_{t,N}^{(1)} = \sum_{k=0}^{\frac{N}{2}-2} c_k^{(N)} \left(\cos\left[\frac{2\pi k t}{N}\right] + \cos\left[\frac{2\pi(k+1)t}{N}\right] \right) + O(\frac{\delta^2}{N}),$$

since the missing summand can be estimated as follows:

$$c_{\frac{N}{2}-1}^{(N)} \left(\cos(t\pi - \frac{2t\pi}{N}) + \cos(t\pi) \right) = (-1)^t c_{\frac{N}{2}-1}^{(N)} (\cos(\pi(1-\delta)) - \cos(\pi)) = O(\frac{\delta^2}{N}).$$

The remaining sum now has an even number of summands, and we can proceed as in the proof of Lemma 10.1 to show the same bound. As for $T_{t,N}^{(2)}$, we observe that (cf. (10.8))

$$T_{t,N}^{(2)} = \sum_{k=0}^{\frac{N}{2}-2} (c_{k+1}^{(N)} - c_k^{(N)}) \cos\left[\frac{2\pi(k+1)t}{N}\right] + (-1)^t (c_{\frac{N}{2}}^{(N)} - c_{\frac{N}{2}-1}^{(N)}),$$

where by definition of $c_k^{(N)}$

$$c_{\frac{N}{2}}^{(N)} - c_{\frac{N}{2}-1}^{(N)} = N\left[\int_0^{\frac{2}{N}} \int_{1-\frac{2}{N}}^1 g(x-y)dxdy - \int_0^{\frac{2}{N}} \int_1^{1+\frac{2}{N}} g(x-y)dxdy\right]$$

$$= N\int_0^{\frac{2}{N}} \int_{1-\frac{2}{N}}^1 \left[g(x-y) - g(x-y+\frac{2}{N})\right]dxdy$$

$$= N\int_0^{\frac{2}{N}} \int_{1-\frac{2}{N}}^{1-\frac{2}{N}+y} \left[g(x-y) - g(x-y+\frac{2}{N})\right]dxdy$$

$$+ N\int_0^{\frac{2}{N}} \int_{1-\frac{2}{N}+y}^1 \left[g(x-y) - g(2-x+y-\frac{2}{N})\right]dxdy,$$

where we used $g(x) = g(2+x) = g(2-x)$. The integrand in the first term satisfies

$$g(x-y) - g(x-y+\frac{2}{N}) = -\frac{2}{N}g'(x-y) - \frac{2}{N^2}g''(\xi(x,y))$$

for some $\xi(x,y) \in [0,1]$, and using $g \in C^2[0,1]$, we easily deduce

$$N\int_0^{\frac{2}{N}} \int_{1-\frac{2}{N}}^{1-\frac{2}{N}+y} \left[g(x-y) - g(x-y+\frac{2}{N})\right]dxdy$$

$$= -2\int_0^{\frac{2}{N}} \int_{1-\frac{2}{N}}^{1-\frac{2}{N}+y} g'(x-y)dxdy + O(\frac{1}{N^3})$$

$$= -2\int_0^{\frac{2}{N}} \int_{1-\frac{2}{N}}^{1-\frac{2}{N}+y} g'(1)dxdy + O(\frac{1}{N^3}) = O(\frac{1}{N^3}),$$

since $g'(1) = 0$. As for the second integral, since $x - y \in [0,1]$, $2 - x + y - \frac{2}{N} \in [0,1]$ for $y \in [0, \frac{2}{N}]$, $x \in [1 - \frac{2}{N} + y, 1]$, we have

$$g(x-y) - g(2-x+y-\frac{2}{N}) =$$

$$= 2(x-y-1+\frac{1}{N})g'(x-y) - 2(x-y-1+\frac{1}{N})^2 g''(\zeta(x,y))$$

for some $\zeta(x,y) \in [0,1]$, so that

$$N\int_0^{\frac{2}{N}} \int_{1-\frac{2}{N}+y}^1 \left[g(x-y) - g(2-x+y-\frac{2}{N})\right]dxdy =$$

$$= 2N \int_0^{\frac{2}{N}} \int_{1-\frac{2}{N}+y}^{1} (x - y - 1 + \frac{1}{N}) g'(x-y) dx dy$$

$$- 2N \int_0^{\frac{2}{N}} \int_{1-\frac{2}{N}+y}^{1} (x - y - 1 + \frac{1}{N})^2 g''(\zeta(x,y)) dx dy$$

$$= 2N \int_0^{\frac{2}{N}} \int_{1-\frac{2}{N}+y}^{1} (x - y - 1 + \frac{1}{N}) g'(x-y) dx dy + O(\frac{1}{N^3})$$

$$= 2N \int_0^{\frac{2}{N}} \int_{1-\frac{2}{N}+y}^{1} (x - y - 1 + \frac{1}{N}) g'(1) dx dy + O(\frac{1}{N^3}) = O(\frac{1}{N^3}),$$

and thus

$$T_{t,N}^{(2)} = \sum_{k=0}^{\frac{N}{2}-2} (c_{k+1}^{(N)} - c_k^{(N)}) \cos\left[\frac{2\pi(k+1)t}{N}\right] + O(\frac{1}{N^3}),$$

and the sum can be rewritten as an "alternating" sum with an even number of summands as in the proof of Lemma 10.3, and the estimation is done analogously up to minor modifications. As for the terms resulting from the matrix D_N (see Lemma 10.5), we have

$$2N \sum_{k=0}^{\frac{N}{2}-1} (-1)^k \int_0^{\frac{1}{N}} x \left[g(\frac{2k+1}{N} - x) - g(\frac{2k+1}{N} + x)\right] dx \left(1 - \cos(\frac{2\pi t}{N})\right)$$

$$= 2N \sum_{k=0}^{\frac{N}{2}-2} (-1)^k \int_0^{\frac{1}{N}} x \left[g(\frac{2k+1}{N} - x) - g(\frac{2k+1}{N} + x)\right] dx \left(1 - \cos(\frac{2\pi t}{N})\right)$$

$$+ 2N \int_0^{\frac{1}{N}} x \left[g(1 - \frac{1}{N} - x) - g(1 - \frac{1}{N} + x)\right] dx \left(1 - \cos(\frac{2\pi t}{N})\right),$$

where

$$g(1 - \frac{1}{N} - x) - g(1 - \frac{1}{N} + x) = -2xg'(1 - \frac{1}{N} + x) - 2x^2 g''(\eta(x))$$

for some $\eta(x) \in [0,1]$. Thus

$$2N \int_0^{\frac{1}{N}} x \left[g(1 - \frac{1}{N} - x) - g(1 - \frac{1}{N} + x)\right] dx =$$

$$= -4N \int_0^{\frac{1}{N}} x^2 g'(1 - \frac{1}{N} + x) dx - 4N \int_0^{\frac{1}{N}} x^3 g''(\eta(x)) dx$$

$$= -4N \int_0^{\frac{1}{N}} x^2 g'(1 - \frac{1}{N} + x) dx + O(\frac{1}{N^3}) = -4N \int_0^{\frac{1}{N}} x^2 g'(1) dx + O(\frac{1}{N^3}) = O(\frac{1}{N^3}),$$

which yields

$$2N \sum_{k=0}^{\frac{N}{2}-1} (-1)^k \int_0^{\frac{1}{N}} x \left[g(\frac{2k+1}{N} - x) - g(\frac{2k+1}{N} + x)\right] dx \left(1 - \cos(\frac{2\pi t}{N})\right)$$

$$= 2N \sum_{k=0}^{\frac{N}{2}-2} (-1)^k \int_0^{\frac{1}{N}} x \left[g\left(\frac{2k+1}{N} - x\right) - g\left(\frac{2k+1}{N} + x\right) \right] dx \left(1 - \cos\left(\frac{2\pi t}{N}\right)\right) + O\left(\frac{1}{N^3}\right).$$

Again, the remaining sum is an alternating one with an even number of summands, so that we can rewrite it as a sum of differences and thus can show Lemma 10.5 for this case in a similar way as the version for multiples of 4.

Bibliography

[1] R. Adams. *Sobolev Spaces*, volume 65 of *Pure and Applied Mathematics, a Series of Monographs and Textbooks*. New York - San Francisco - London: Academic Press, Inc., a subsidiary of Harcourt Brace Jovanovich, Publishers., 1975.

[2] G. Alberti. Variational models for phase transitions, an approach via Γ-convergence. In *Buttazzo, G. (ed.) et al.: Calculus of variations and partial differential equations. Topics on geometrical evolution problems and degree theory. Based on a summer school, Pisa, Italy, September 1996*, pages 95–114, 327–337. Berlin: Springer-Verlag, 2000.

[3] G. Alberti and G. Bellettini. A non-local anisotropic model for phase transitions: Asymptotic behaviour of rescaled energies. *European Journal of Applied Mathematics*, 9(3):261–284, 1998.

[4] G. Alberti and G. Bellettini. A nonlocal anisotropic model for phase transitions. *Mathematische Annalen*, 310(3):527–560, 1998.

[5] G. Alberti, G. Bouchitté, and P. Seppecher. Un résultat de perturbations singulières avec la norme $H^{1/2}$. *Comptes Rendus de l'Académie des Sciences, Paris, Series I*, 319(4):333–338, 1994.

[6] G. Alberti and S. Müller. A new approach to variational problems with multiple scales. *Communications on Pure and Applied Mathematics*, 54(7):761–825, 2001.

[7] J.M. Ball. A version of the fundamental theorem for Young measures. In *Rascle, Michel (ed.) et al.: PDEs and Continuum Models of Phase Transitions. Proceedings of an NSF-CNRS Joint Seminar held in Nice, France, January 18-22, 1988*, volume 344 of *Lecture Notes in Physics*, pages 207–215. Berlin etc.: Springer-Verlag, 1989.

[8] J.M. Ball and R.D. James. Fine phase mixtures as minimizers of energy. *Archive for Rational Mechanics and Analysis*, 100:13–52, 1987.

[9] J.M. Ball and R.D. James. Proposed experimental tests of a theory of fine microstructure and the two-well problem. *Philosophical Transactions of the Royal Society of London. Series A*, 338:389–450, 1992.

[10] H. Berliocchi and J.-M. Lasry. Intégrandes normales et mesures paramètrées en calcul des variations. *Bulletin de la Société Mathématique de France*, 101:129–184, 1973.

[11] J.W. Cahn and J.E. Hilliard. Free energy of a nonuniform system I: Interfacial free energy. *Journal of Chemical Physics*, 28:258–267, 1958.

[12] A. Chmaj and X. Ren. Multiple layered solutions of the nonlocal bistable equation. *Physica D*, 147(1-2):135–154, 2000.

[13] A. Chmaj and X. Ren. The nonlocal bistable equation: Stationary solutions on a bounded interval. *Electronic Journal of Differential Equations*, 2002(Paper No. 02):1–12, 2002.

[14] B. Dacorogna. *Direct Methods in the Calculus of Variations*, volume 78 of *Applied Mathematical Sciences*. Berlin etc.: Springer-Verlag, 1989.

[15] D. Dal Maso. *An Introduction to Γ-Convergence*, volume 8 of *Progress in Nonlinear Differential Equations and their Applications*. Basel: Birkhäuser, 1993.

[16] P. Davis. *Circulant Matrices*. Pure and applied Mathematics. A Wiley-Interscience Publication. New York etc.: John Wiley & Sons, 1979.

[17] E. De Giorgi. Convergence problems for functionals and operators. In *Recent Methods in Non-linear Analysis, Proc. Int. Meet., Rome 1978*, pages 131–188. Bologna: Pitagora, 1979.

[18] E. De Giorgi and T. Franzoni. Su un tipo di convergenza variazionale. *Atti della Accademia Nazionale dei Lincei, VIII. Serie, Rendiconti, Classe di Scienze Fisiche, Matematiche e Naturali*, 58:842–850, 1975.

[19] P. Delsarte and Y. Genin. Spectral properties of finite Toeplitz matrices. In *Mathematical theory of networks and systems, Proc. int. Symp., Beer Sheva/Isr. 1983*, volume 58 of *Lecture Notes in Control and Information Sciences*, pages 194–213. Berlin: Springer-Verlag, 1984.

[20] A. DeSimone, R.V. Kohn, S. Müller, and F. Otto. A reduced theory for thin-film micromagnetics. *Communications on Pure and Applied Mathematics*, 55(11):1408–1460, 2002.

[21] R.E. Edwards. *Functional Analysis. Theory and Applications*. New York-Chicago-San Francisco-Toronto-London: Holt Rinehart and Winston, 1965.

[22] J. Elstrodt. *Maß- und Integrationstheorie*. Springer-Lehrbuch. Berlin: Springer-Verlag, 1996.

[23] L.C. Evans and R.F. Gariepy. *Measure Theory and Fine Properties of Functions*. Studies in Advanced Mathematics. Boca Raton: CRC Press., 1992.

[24] D. Gilbarg and N.S. Trudinger. *Elliptic Partial Differential Equations of Second Order. 2nd Ed.*, volume 224 of *Grundlehren der Mathematischen Wissenschaften*. Berlin etc.: Springer-Verlag, 1983.

[25] R.M. Gray. Toeplitz and circulant matrices: A review. http://ee.stanford.edu/ ~gray/toeplitz.pdf.

[26] J. Hale. *Ordinary Differential Equations. 2nd Ed.*, volume 21 of *Pure and Applied Mathematics*. Orig. publ. by Wiley-Interscience. Huntington, New York: Robert E. Krieger Publishing Company, 1980.

[27] A.G. Khachaturyan. Some questions concerning the theory of phase transformations in solids. *Soviet Physics Solid State*, 8:2163–2168, 1967.

[28] A.G. Khachaturyan and G.A. Shatalov. Theory of macroscopic periodicity for a phase transition in the solid state. *Soviet Physics JETP*, 29:557–561, 1969.

[29] R.V. Kohn and S. Müller. Branching of twins near an austenite - finely-twinned-martensite interface. *Philosophical Magazine A*, 66(5):697–715, 1992.

[30] R.V. Kohn and S. Müller. Surface energy and microstructure in coherent phase transitions. *Communications on Pure and Applied Mathematics*, 47(4):405–435, 1994.

[31] R.V. Kohn and P. Sternberg. Local minimisers and singular perturbations. *Proceedings of the Royal Society of Edinburgh, Section A*, 111(1/2):69–84, 1989.

[32] J.L. Lions and E. Magenes. *Non-Homogeneous Boundary Value Problems and Applications. Vol. I. Translated from the French by P. Kenneth*, volume 181 of *Grundlehren der mathematischen Wissenschaften*. Berlin-Heidelberg-New York: Springer-Verlag, 1972.

[33] L. Modica. The gradient theory of phase transitions and the minimal interface criterion. *Archive for Rational Mechanics and Analysis*, 98:123–142, 1987.

[34] L. Modica and S. Mortola. Il limite nella Γ-convergenza di una famiglia di funzionali ellittici. *Bollettino della Unione Matematica Italiana, V. Ser., A*, 14:526–529, 1977.

[35] L. Modica and S. Mortola. Un esempio di Γ⁻-convergenza. *Bollettino della Unione Matematica Italiana, V. Ser., B*, 14:285–299, 1977.

[36] S. Müller. Minimizing sequences for nonconvex functionals, phase transitions and singular perturbations. In *Problems Involving Change of Type, Proc. Conf., Stuttgart/FRG 1988*, volume 359 of *Lecture Notes in Physics*, pages 31–44. Berlin etc.: Springer-Verlag, 1990.

[37] S. Müller. Singular perturbations as a selection criterion for periodic minimizing sequences. *Calculus of Variations and Partial Differential Equations*, 1(2):169–204, 1993.

[38] S. Müller. Variational models for microstructure and phase transisions. In *Hildebrandt, S. (ed.) et al., Calculus of variations and geometric evolution problems. Lectures given at the 2nd session of the Centro Internazionale Matematico Estivo (CIME), Cetraro, Italy, June 15-22, 1996*, volume 1713 of *Lecture Notes in Mathematics*, pages 85–210. Berlin: Springer-Verlag, 1999.

[39] Y. Nishiura and I. Ohnishi. Some mathematical aspects of micro-phase separation in diblock copolymers. *Physica D*, 84:31–39, 1995.

[40] T. Ohta and K. Kawasaki. Equilibrium morphology of block copolymer melts. *Macromolecules*, 19:2621–2632, 1986.

[41] X. Ren and J. Wei. On the multiplicity of solutions of two nonlocal variational problems. *SIAM Journal of Mathematical Analysis*, 31(4):909–924, 2000.

[42] X. Ren and J. Wei. Concentrically layered energy equilibria of the di-block copolymer problem. *European Journal of Applied Mathematics*, 13(5):479–496, 2002.

[43] X. Ren and J. Wei. On energy minimizers of the di-block copolymer problem. *Interfaces and Free Boundaries*, 5:193–238, 2003.

[44] A.L. Roitburd. The domain structure of crystals formed in the solid phase. *Soviet Physics Solid State*, 10:2870–2876, 1969.

[45] M. Valadier. A course on Young measures. *Rendiconti dell'Istituto di Matematica dell'Universita di Trieste*, 26:349–394, 1994. suppl.

[46] R.S. Varga. Eigenvalues of circulant matrices. *Pacific Journal of Mathematics*, 4:151–160, 1954.

[47] L.C. Young. *Lectures On the Calculus of Variations and Optimal Control Theory. 2nd Edition*. New York: Chelsea Publishing Company, 1980.

[48] E. Zeidler. *Nonlinear Functional Analysis and its Applications. III: Variational Methods and Optimization*. New York etc.: Springer-Verlag, 1985.

Index